Springer Theses

Recognizing Outstanding Ph.D. Research

Aims and Scope

The series "Springer Theses" brings together a selection of the very best Ph.D. theses from around the world and across the physical sciences. Nominated and endorsed by two recognized specialists, each published volume has been selected for its scientific excellence and the high impact of its contents for the pertinent field of research. For greater accessibility to non-specialists, the published versions include an extended introduction, as well as a foreword by the student's supervisor explaining the special relevance of the work for the field. As a whole, the series will provide a valuable resource both for newcomers to the research fields described, and for other scientists seeking detailed background information on special questions. Finally, it provides an accredited documentation of the valuable contributions made by today's younger generation of scientists.

Theses are accepted into the series by invited nomination only and must fulfill all of the following criteria

- They must be written in good English.
- The topic should fall within the confines of Chemistry, Physics, Earth Sciences, Engineering and related interdisciplinary fields such as Materials, Nanoscience, Chemical Engineering, Complex Systems and Biophysics.
- The work reported in the thesis must represent a significant scientific advance.
- If the thesis includes previously published material, permission to reproduce this must be gained from the respective copyright holder.
- They must have been examined and passed during the 12 months prior to nomination.
- Each thesis should include a foreword by the supervisor outlining the significance of its content.
- The theses should have a clearly defined structure including an introduction accessible to scientists not expert in that particular field.

More information about this series at http://www.springer.com/series/8790

Alistair Inglis

Investigating a Phase Conjugate Mirror for Magnon-Based Computing

Doctoral Thesis accepted by
University of Oxford, UK

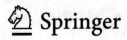 Springer

Author
Dr. Alistair Inglis
Clarendon Laboratory
University of Oxford
Oxford, UK

Supervisor
Prof. John F. Gregg
Clarendon Laboratory
University of Oxford
Oxford, UK

ISSN 2190-5053   ISSN 2190-5061   (electronic)
Springer Theses
ISBN 978-3-030-49747-7   ISBN 978-3-030-49745-3   (eBook)
https://doi.org/10.1007/978-3-030-49745-3

This Springer imprint is published by the registered company Springer Nature Switzerland AG
The registered company address is: Gewerbestrasse 11, 6330 Cham, Switzerland

Supervisor's Foreword

Anatole Abragam, author of "The Principles of Nuclear Magnetism", was wont to joke with colleagues that there was little that a family of spins would not do to oblige, provided that one spoke with them nicely. He was referring to nuclear spins, but in recent years, his lighthearted words have been soundly vindicated in a more general sense by the spectacular advances in the field of Spintronics that is known as Magnonics.

Spin waves, whose quanta are called magnons, first made an entry into scientific vocabulary in the late 1940s when James Griffiths used the 2 centimetre radar technology—that was developed for wartime aircraft detection—to examine ferromagnetic metals, and he was the first to observe ferromagnetic resonance: spin waves with zero k-vector. However, the full flowering of magnonic science had to wait more than half a century for the advent of ultrafast electronics which enables these extraordinary wave phenomena to be examined "up close and in personal".

Classic magnetism textbooks focus exclusively on very short wavelength, high-frequency spin waves and ascribe to them a quadratic dispersion relation that derives from assuming a restoring force that is exchange-based and acts exclusively between adjacent spins. However, at lower frequencies where the exchange interaction "plays second fiddle" to long-range magnetostatic interaction between spins, many atomic spacings apart, new and unusual behaviour gives rise to a wealth of intriguing new physics. When bulk material is exchanged for thin film magnetic samples, this novel behaviour becomes even more extreme.

A striking feature of these lower frequency magnetostatic spin waves in thin films is the appearance of dispersion relations with negative group velocity that gives rise to such mind-bending phenomena as Inverse Doppler Effect. Moreover, these various magnon dispersion relations may be displaced in frequency by applying local magnetic bias fields which allow for the creation of a novel type of artificial crystal whose mesoscopic periodicity may be switched on and off in real time; and this in turn has been used, among other applications, for the creation of a time-reversal signal processor. Magnonic switches, transistors and digital logic gates that function using wave computing have been constructed and demonstrated. The most thought-provoking of the latter is a majority gate that is re-programmable "on the fly" like a Field Programmable Gate Array but that is additionally capable of applying different signal processing protocols to multiple parallel datastreams simultaneously—a task that no current silicon device can emulate. Herein lies the promise even of new computing paradigms that exploit this unprecedented architecture.

All of this spectacular new signal processing power has the potential to transform the future of information technology. Laptop clock speeds have been frozen for almost two decades as an emergency measure to prevent integrated circuit heat death; and computer server farms are starting to rival the petrol engine as a major contributor to global warming. Magnonic computing devices—both analog and digital—run at one thousandth of the power of their silicon equivalents, occupy similar real estate and have an ultimate speed limit in the TeraHertz, so they are strategically placed to ward off the threatened apocalypse by which Moore's Law hits the buffers.

An aspect of the low power operation of thin film magnonics is the ability of spin waves to exhibit strong nonlinearity at very low power levels. Much exciting physics lies in the domain of non-linear optics, but the laser powers—and footprints —required to explore this science are enormous. In magnonics, not only can extreme nonlinearity be induced at microwatt levels, but the effect of this nonlinearity can be carefully controlled by bandwidth considerations so that it appears only when needed.

In this work, the author uses a magnonic system as the theatre for a piece of non-linear physics that invokes optical parallels to create a four-wave mixing device capable of signal phase conjugation. In so doing, he not only delivers yet another practical analog computing device to add to the armoury of magnonic computing hardware, but also demonstrates the utility, the versatility and the elegance of this rapidly growing new field of solid state physics.

Oxford, UK   John F. Gregg

Abstract

This thesis reports on the realisation of a phase conjugate spin wave device employing a four-wave mixing process. Unexpected nonlinear behaviour was revealed in further experiments which are reported on latterly.

Chapter 1 introduces the motivation for this thesis: the need to avoid the stagnation of technological progress by investigating alternative computing paradigms, namely magnon-based computing. An introduction to the field of magnonics and magnon behaviour is presented, followed by a brief introduction to nonlinear behaviour.

Chapter 2 builds on the general description of spin waves and introduces the concept of phase conjugation. Following an optical treatment of phase conjugation via four-wave mixing, a theoretical description of the phase conjugation in the spin wave regime is presented.

Chapter 3 details the experimental materials and methods employed in the succeeding chapters. A description of the antenna design and fabrication processes is presented along with a description of the key experimental equipment.

Chapter 4 describes an experimental investigation concerning the realisation of a phase conjugate mirror via four-wave mixing created in a spin wave system. It is demonstrated through experiments and simulations that the mirror is most reflective when a standing spin wave is present across the width of the magnon waveguide.

Chapter 5 investigates a time-domain fractal arising from the spatio-temporally periodic potential that occurs as a result of the standing wave. The onset of great-granddaughter fractal modes is observed and the frequency dependence of the amplitude of the fractal modes is discussed.

Chapter 6 offers concluding remarks, and considers future experiments that may develop the field.

Acknowledgements

Thanks to my supervisor Prof. John Gregg for rolling up his sleeves and leading by example. His physical insight, good humour, and words of encouragement are always appreciated.

I would also like to thank Dr. Alexy Karenowska for general magnonic discussion and providing vital materials, Paul Pattinson for advice on photofabrication, and Mathew Newport for his assistance in the research workshop.

For hosting me at the University of Kaiserslautern, I would like to thank Prof. Burkard Hillebrands, Dr. Philipp Pirro, Dr. Dymtro Bhozko, and Moritz Geilen.

For creating a hugely enjoyable working environment, I wish to thank my lab mates and fellow spin wave enthusiasts: Dr. Nelson Fung, Dr. Calvin Tock, Dr. Richard Morris, Dr. Sandoko Kosen, Dr. Arjan Van Loo, and Aaron Briggs.

For always being there to remind me of the joys of humanity, I must thank a kind few: Kathleen, James, Jay, Andre, Rob, Ethan, Hamish, John, Faith, Derek, and Kathy.

An honourable mention goes to my family, in particular my parents, without whom I would quite literally be nothing. Thank you, Hazel, Matt, Fiona, and Carolyn.

Thank you to Magdalen College and the late Iris and Leon Beghian for their generosity in funding this research.

June 2019   Alistair Inglis

Contents

Chapter 1
Motivation and Introduction to Theory

The question now is what will happen in the early 2020s, when continued scaling is no longer possible with silicon because quantum effects have come into play. What comes next?

Mitchell Waldrop More than Moore [1]

Broadly speaking, this thesis is concerned with the study of nonlinear magnon processes that may be exploited in novel wave-computing based devices. This introductory chapter explains the relevance of such devices, placing them in the context of magnon-based computing, and the wider field of wave computing in general. Following this motivational foundation, essential elements of the theory of magnons are presented in preparation for topics discussed in later chapters. Latterly, a brief introduction to nonlinear magnon behaviour is set out.

1.1 Magnonic Computing

Material possessing long range magnetic order—the media with which this thesis is concerned—have been of technological importance for almost 1000 years [2]. One of the greatest technological developments of the last century was the invention of magnetic storage, allowing for the permanent and compact storage of information in the form of binary 'bits'. The 1990s saw the advent of spintronics with the observation of spin-dependent electronic transport [3, 4], opening the door for information transport via angular momentum flow. With this came a pathway by which one could use spin waves to carry information, rather than ballistic spin currents. Given the unimaginable success of semiconductor transistor computing however, do we need a new technology at all?

© The Editor(s) (if applicable) and The Author(s), under exclusive license
to Springer Nature Switzerland AG 2020
A. Inglis, *Investigating a Phase Conjugate Mirror for Magnon-Based Computing*,
Springer Theses, https://doi.org/10.1007/978-3-030-49745-3_1

1.1.1 The Problem to be Solved

The history of complementary semiconductor-metal-oxide (CMOS) technology is one of exponential improvement. The complexity and abundance of devices has exploded from the few crude home-computers of the 1970s to the billions of smart-phones that dominate modern life. The enduring supremacy of CMOS technology is rooted in the scaling trends that have been enjoyed for decades [1]. Transistor-size scaling is widely associated with two trends: Moore's law [5] and Dennard's scaling [6]. The former, while referred to as a law, is merely an observation that the number of transistors per integrated circuit roughly doubles every two years. The long-term industrial adherence to this trend however, is leading to an inevitable, long heralded dead end [7–9].

As the size of commerically produced transistors has reached 7 nm [10], the power required to ensure reliable switching has deviated significantly [11] from Dennard's scaling which states that the power density of a circuit will stay approximately con-stant as transistors become smaller. Keeping devices at manageable power levels is of paramount concern in an age of portable, 'smart' devices. As transistors continue to shrink however, the power required to combat uncertainties associated with quantum effects will become non-viable and alternative forms of computation will become ever more appealing.

Scaling and power consumption are by no means the only challenges facing CMOS computing. Since 2004 [1] the increase in processor clock speeds has effec-tively stalled, with rates plateauing at approximately 3 GHz. The physical limitations imposing this stasis lie in the problem of heat dissipation. To increase clock speeds would be to increase the rate of Joule-heating, leading to devices running at temper-atures unfit for purpose.

A number of potential alternative devices have been identified for CMOS replace-ment [12]. The focus of this thesis relies on one technology in particular: spin wave computing, or magnonics. Spin wave devices are particularly appealing in light of the obstacles facing the progress of CMOS discussed above.

In a wave computing paradigm, the wave frequency defines the clock speed [13]. The magnon spectrum spans from the gigahertz region, where current telecommu-nications technology is located, up into the terahertz region where there is increased scientific and technological interest [14].[1] The implementation of a magnon based computing system would therefore unlock higher clock speeds beyond the limits of CMOS.

In tandem with an increased clock speed, is a reduction in component size. The minimum size is defined by the wavelength of the spin wave, the lower bound of which is set by the lattice constant of the spin wave medium. Since for exchange magnons, the frequency increases quadratically with wavelength, these two beneficial aspects of magnon-based computing—nanometer scalability and increased clock speeds—go

---

[1] See also, for example the recent Special Topic section devoted to "Advances in Terahertz Solid-State Physics and Devices" published in the Journal of Applied Physics **125**(15) (2019).

hand in hand. Furthermore, spin wave devices have been shown to have significantly lower power consumption than 10 nm CMOS equivalent devices [15].

In discussing a few benefits of magnon-based computing we have offered wave-based solutions to CMOS-based problems. There is an important feature of wave computing however, that solves a problem that is not immediately facing CMOS, yet offers an improvement nonetheless. A superior aspect of wave computing, not limited to magnonics, is the ability to control—and therefore encode information into—two parameters: amplitude and phase. This extra parameter constitutes an extra 'bit', and we shall now turn our attention to how these may be exploited.

1.1.2 Spin Wave Devices

Over recent years there has been a growing number of spin wave devices and logic gates [13]. An early, and relatively simple example is the Mach–Zender interferometer [16, 17], a schematic of which is shown in Fig. 1.1. We can see from the figure that this spin wave device has the same functionality as a XNOR logic gate. Two magnon waveguides each have a current-carrying wire placed over them which act as the logical inputs. When current flows the input is set to a '1' and will introduce a π phase shift along the waveguide. Therefore, if both inputs have the same value, each spin wave channel will have the same phase shift and interfere constructively, resulting in a logical '1' output, while if only one input is set to '1', then the channels will destructively interfere, returning a logical '0' at the output.

Another spin wave device operating in accordance with Boolean logic is the majority gate, depicted in Fig. 1.2. This three-input device accepts binary data encoded in the phase of the input magnons (relative to some reference signal), wherein a phase offset of 0 would correspond to a logical '0', and a phase difference of π would correspond to a logical '1'. The output is simply the confluence of the input channels, where the phase of the output is equal to the majority of the input phases. The novelty of this device lies in its ability to behave as four traditional logic gates: AND, OR, NAND, and NOR. As shown in the truth table in Fig. 1.2, by treating input 3 as a control-bit and fixing it at either 0 or 1, the gate behaves as an AND or OR gate, respectively. Furthermore, since the information is encoded in the phase of the wave, by moving the read-out by one half-cycle, the output will behave as either NAND or NOR. Due to the versatility of the majority gate, it can in some cases offer substantially reduced architecture-footprints. To assemble a full-adder circuit for example would require 28 transistors in CMOS, but would only require 3 majority gates in a magnon-based paradigm.

A major drawback of this technology, exemplified by the logic gates described above, is that they cannot be used to drive one another without the use of some magneto-electrical interconversion method. The relatively weak conversion efficiencies associated with this process, fuelled the desire for the means to control one magnon current with another magnon current. The search for such a device would

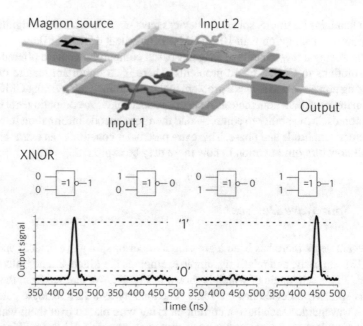

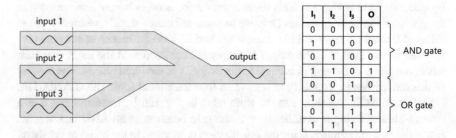

Fig. 1.1 Schematic of Mach Zender interferometer. A coherent source excites magnons in identical waveguides, each with a current-carrying wire placed over them, acting as a logical input. When current is flowing in a wire, the logical input is '1' and the locally distorted magnetic field introduces a phase shift of π in the waveguide. The constructive or destructive interference of the magnons at the output translates into a logical '1' or '0', respectively, corresponding to the behaviour of an XNOR gate. Image taken from Ref. [13]

I_1	I_2	I_3	O	
0	0	0	0	
1	0	0	0	
0	1	0	0	AND gate
1	1	0	1	
0	0	1	0	
1	0	1	1	
0	1	1	1	OR gate
1	1	1	1	

Fig. 1.2 Sketch and truth table describing the magnon majority gate. The device (left) shows three inputs. Inputs 1 and 2 are in phase with each other and have a phase shift of 0 radians with respect to some reference signal. The binary encoding of these inputs is '0', while input 3 is π radians out of phase and has a binary value of '1'. The output is simply the superposition of the inputs, which after interference leaves an output with a 0 radians phase shift and logical value '0', which is the majority value of the inputs. The truth table shows that if input 3 is used as a control bit, the behaviour of the gate can change between AND and OR

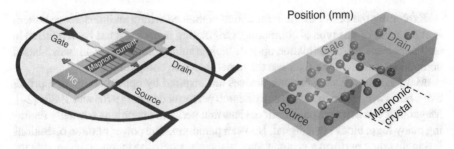

Fig. 1.3 Illustrations describing the magnon transistor. Carrier magnons (blue spheres) flow from source to drain. Gate magnons (red spheres) get trapped in the magnonic crystal and nonlinearly scatter the carrier magnons, resulting in a reduced magnon current at the drain. Images taken from Ref. [18]

also satisfy one of the five tenets of computing; that the output of one device can drive the next one.[2]

The breakthrough came in 2014 with the advent of the magnon transistor [18], depicted in Fig. 1.3. This three terminal device enables the control of the magnon current density flowing from the source to the drain (blue spheres in figure), by the injection of gate magnons (red spheres). The gate magnons are confined by the magnonic crystal [19], and scatter the source-drain magnon current nonlinearly. The greater the density of gate magnons, the more scattering experienced by the source-drain current and the more the gate closes. The device has been shown to decrease the source-drain current by up to three orders of magnitude by controlling the density of gate magnons. This particular device describes just one method for controlling magnon flow; other architectures have recently been proposed [20, 21].

1.1.2.1 Non-Boolean Spin Wave Devices

The Mach–Zender interferometer and the majority gate described above give examples of Boolean logic operations using the two key parameters in wave computing: amplitude and phase. These devices require that the controlling parameters are digitised into binary logic. While progress on developing *digital* spin wave architecture is ongoing [22, 23], there is a resurgence of interest in exploiting the continuous nature of the phase and amplitude to realise a type of *analogue* computing.

---

[2]The 'logic tenets' as described in Ref. [12] are (1) nonlinear device characteristics (related to acceptable noise margin), (2) complete set of Boolean operators, (3) one device output can drive another device (4) power amplification, or gain >1, and (5) feedback prevention, where the output does not effect the input. (1) and (2) certainly are satisfied by the logic gates described above, as is (3) by the magnon transistor. Conditions (4) and (5) however are tricky and currently require some electrical input, which, in any practical system would be undesirable due to the poor magnon-voltage interconversion efficiency.

Exploiting the power of the information contained within an interference pattern is at the heart of this type of computing. Originating in optics, it has been referred in as 'Fourier optics', 'information optics', 'holographic computing', and 'wave-based computing',[3] though we shall use the term 'analogue computing'.

Many non-Boolean spin wave devices are inspired by optical computing primitives, borrowing from the extensive research performed in the 1970s and 1980s [24]. Magnonics can offer purpose-built devices that perform operations normally requiring many logic blocks in a digital, Boolean paradigm. The power of these operations lies in that they perform a vector-vector mapping, where the mapping represents the computational task.

An incredibly useful non-Boolean process is the analogue Fourier transform. Optically, this can be achieved using a simple convex lens. Recently, micromagnetic simulations [25] and experimental work [26] have demonstrated that with the implementation of a concave effective magnetic field distribution, one can create a spin wave lens to perform the same Fourier transform process. Such a device would in principle enable one to perform Fourier transforms in a single step, at a rate of many gigasamples per second, far beyond current CMOS capabilities [27].

Recent work also shows the viability of another analogue spin wave device: the spectrum analyser [28]. In this passive device, based on a Rowland circle, a microwave signal excites a spin wave that propagates through a patterned film, forming an interference pattern in the film. The intensity distribution that is obtained corresponds to a spectral decomposition of the original signal (Fig. 1.4).

The use of magnonic crystals [31, 32]—magnetic media that have an artificial spatially periodic variation of their magnetic properites—is a widespread method for altering magnon behaviour to realise nonlinear effects that may be exploited in a computational setting. The periodic modification is usually static, as with a periodic etching or patterning. The ability to apply the periodic alterations to a crystal on a timescale shorter than the propagation time of a spin waves across it however, results in a novel physical effect [33]. This is achieved by placing a meander structure current-carrying wire in close proximity to a magnon waveguide. The periodicity of the meander structure, imposes a periodic perturbation to the magnetic field of the waveguide which can be turned on and off by controlling the current. Using this device, time-reversal of a pulse train was achieved when the perturbation was turned on at a time when the pulse was fully within the periodic structure [30]. The digitally complex task of time-reversal may be performed near-instantaneously by this analogue device.

---

[3] We shall reserve the term 'wave-based computing' or 'wave computing' for any process in which information is stored and/or processed in a wave. This umbrella term encapsulates within it the devices that apply Boolean logic and binary encoding to waves, where as 'Fourier optics' and 'analogue computing' are by their nature distinct from these digital devices. When relying on the continuous aspect of the control parameters—phase and amplitude—we shall use the term 'analogue computing'.

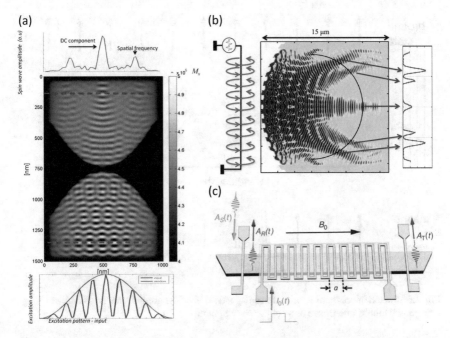

Fig. 1.4 Some examples of analogue computing spin wave devices. The spin wave lens performing a Fourier transform is shown in (**a**), image taken from Ref. [29]. **b** Shows the design for a spectrum analyser, based on a Rowland circle. Image from [25]. The dynamic magnonic crystal, leading to time-reversal and frequency inversion is shown in (**c**). Figure taken from Ref. [30]

Central to this thesis is the realisation of a magnonic device that can perform an analogue computing operation on the phase of a spin wave, and the nonlinear behaviour that arises from such a device. Before the experimental details may be presented however, an introduction to the physics behind magnons must be discussed.

1.2 What is a Magnon?

A magnon is a quasi-particle associated with the quantisation of a spin wave. Magnons represent an excitation from the fully ordered magnetic ground state of a system. The notion of a magnon may be easily conceptualised by considering the example of a simple 1D ferromagnetic chain. When in its ground state, all of the spins are aligned in the same direction. As energy enters the system, a deviation from the ground state occurs which, in the spirit of wave-particle duality, may be considered in two ways.

The 'particle-like' magnon is a delocalised spin-flip that can move along the chain. The 'wave-like' description can be thought of more classically: as energy enters the system, the spins can afford to no longer align perfectly with the magnetic field. This

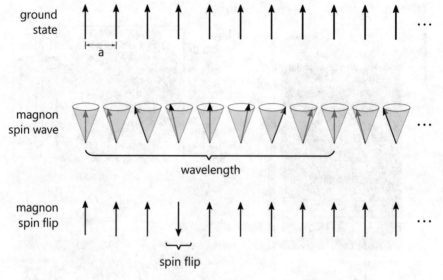

Fig. 1.5 Sketch of magnons on a 1D ferromagnetic chain. The ground state is excited, resulting in a magnon which is interpreted as a spin wave or spin flip

canting however, causes each spin to precess about the magnetic field. The phase of the precession propagates between spins in the chain, where the phase difference between neighbouring spins defines the wavelength, as illustrated in Fig. 1.5. Throughout this thesis we will be working within the wave-like interpretation.[4]

Typically, a physicist's first introduction to magnons is an undergraduate treatment of exchange spin waves, so-called because the dominant coupling term between spins is the exchange interaction. These waves are usually treated as an introduction to magnetisation dynamics because the dispersion relation for exchange spin waves may be derived classically for a 1D ferromagnetic chain of N spins (as in Fig. 1.5), starting from a nearest-neighbour coupling between spins for the form:

$$U = -2J \sum_{p=1}^{N} \mathbf{S}_p \cdot \mathbf{S}_{p+1}, \tag{1.1}$$

where J is the exchange integral and $\hbar \mathbf{S}_p$ is the angular momentum of the spin at the pth site. Equation (1.1) can be rewritten in terms of the magnetic moment and magnetic field at site p, denoted by $\boldsymbol{\mu}_p$, and $\mathbf{B}_p$, respectively. Spin dynamics are then introduced using the magnetic torque equations wherein the rate of change of

---

[4]A general introduction to magnons and detailed discussion of their physical origin, can be found in most condensed matter physics undergraduate textbooks. Particularly good examples are Kittel's *Introduction to Solid State Physics* [34] and *Solid State Physics* by Ashcroft and Mermin [35]

angular momentum $\hbar \mathbf{S}_p$ is equal to the torque $\boldsymbol{\mu}_p \times \mathbf{B}_p$. After assuming a wave-like solution and some simple algebra, the following dispersion relation is arrived at:

$$\omega = \frac{2J}{\hbar}\big[1 - (ka)\big] \tag{1.2}$$

$$\omega \approx \frac{2Ja^2}{\hbar}k^2, \tag{1.3}$$

where it has been assumed that ka is small. Here, ω is the magnon angular frequency, k is the magnon wavenumber, and a is the distance between neighbouring spins.

At short lengthscales, the exchange coupling is strong and is the dominant coupling between spins. Since the energy term is so large, so too are the frequencies and the group velocities of the associated magnons. Despite the desirability of these qualities, there is very little literature on exchange magnons in a wave computing setting. This is largely due to the experimental challenges of exciting such small wavelengths, though recently wavelengths down to 125 nm were reportedly excited [36], while the observation of magnon wavelengths as little as 40 nm has also been suggested [37]. These results, though valuable, are very much atypical when viewed against the wealth of literature that concerns magnonic devices that utilise the longer wavelength dipolar magnons.

Magnon behaviour is defined by the two mechanisms by which the spins of ferri- and ferromagnetic lattices are coupled; the exchange coupling (discussed above) which is strong over a short range, and the dipole-dipole coupling, which is longer range and relatively weak. The balance of these two interactions over different length scales leads to a rich and varied dispersion that has three characteristic regimes, shown in Fig. 1.6.

At large k, the exchange coupling dominates, and the quadratic dependence $\omega \propto k^2$ of Eq. (1.3) becomes plain. At small values of k, the dipole-dipole coupling term dominates the total magnetic energy associated with the excitation and spin waves propagate in the dipolar regime. At intermediate lengthscales, both terms contribute to the excitations and a region of dipole-exchange spin waves completes the dispersion curves.

The dipole-dipole interaction between the spins of bound electrons, is of a significantly lower energy than the exchange interaction [39]. The popularity of dipolar magnons in magnonics research is largely due to the ease with which spin waves of this energy (low GHz) and wavelength (order 100 μm) may be excited experimentally, the details of which are discussed in Sect. 3.2. The strongly anisotropic nature of the dipole-dipole interaction translates into rich and varied dispersion relations for spin waves in a magnetic thin film, highly dependent on the orientation of the applied magnetic field with respect to the normal of the film, as we shall discuss in Sect. 1.3.

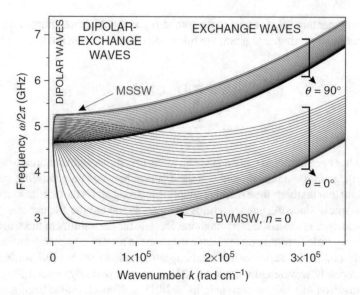

Fig. 1.6 Graph depicting spin wave dispersion in a yttrium iron garnet film of thickness $5\,\mu m$ and saturation magnetisation $M_{sat} = 140\,kA\,m^{-1}$ with an applied magnetic field of $H_0 = 1000\,Oe$. The three characteristic wavelength-dependent regimes are depicted: dipolar, dipolar-exchange, and exchange spin waves. θ corresponds to the angle between the applied field and the spin wave propagation, while n refers to the thickness mode number, with 25 modes shown. The red line refers to the magnetostatic surface spin wave (MSSW), and the blue line the lowest order backward volume magnetostatic spin wave (BVMSW). Image is taken from Ref. [38]

1.3 Dipolar Spin Waves in Magnetic Films

We now turn our attention to the dispersion of dipolar spin waves; the type of magnons employed throughout this thesis. Due to the strong dependence on waveguide geometry, the focus will be restricted to spin waves in magnetic films that have been magnetised to saturation by an external magnetic field. In the dipolar regime the orientation of magnon wavevector, the effective magnetic field, and the normal of the supporting waveguide strongly influences the dispersion characteristics of the spin wave.

When the field is parallel to the film normal and perpendicular to the direction of spin wave propagation, the system has a symmetry that leads to an isotropic dispersion. The waves that are excited in this case are forward volume magnetostatic spin waves (FVMSWs). When the field is in plane however, the symmetry is broken and the dispersion is less straight forward. With the field in the plane of the film—depending on the angle between the field and the wavevector—a backwards volume magnetostatic spin wave (BVMSW) or a magnetostatic surface spin wave (MSSW) may be excited. These types of dipolar spin wave are depicted in Fig. 1.6 where the different dispersion characteristics are clearly visible. In what follows, each of the three types of dipolar spin wave modes are discussed.

The goal of this section will be to obtain expressions relating the frequency to the wavelength for the type of dipolar magnons that are at the core of this thesis. We begin by introducing an equation of motion for the magnetisation and then using it to obtain a description of how a magnetic material responds to a small time-varying magnetic field. From there, the magnetostatic approximation is made, which leads to Walker's equation. When this equation is solved for various geometries, the three different dipolar dispersion relations are obtained and presented.

1.3.1 Polder Susceptibility

The equation of motion for the magnetisation is the well known Landau-Lifschitz equation. For the sake of clarity, we shall use the lossless form:

$$\frac{d\mathbf{M}}{dt} = \gamma \mathbf{M} \times \mathbf{H}_{\text{eff}}, \tag{1.4}$$

where $\mathbf{M}$ is the magnetisation, γ is the gyromagnetic ratio, and $\mathbf{H}_{\text{eff}}$ is the effective magnetic field. The effective magnetic field is the sum of all the torque-producing contributions and can include the effective field arising from exchange, anisotropy, Dzyaloshinskii–Moriya interactions, etc. We shall however, consider an effective field that can be split into a large static contribution and a small time-varying contribution:

$$\mathbf{H}_{\text{eff}} = \mathbf{H}_0 + \mathbf{h}(t), \tag{1.5}$$

and similarly,

$$\mathbf{M} = \mathbf{M}_0 + \mathbf{m}(t). \tag{1.6}$$

Expanding Eq. (1.4) using Eqs. (1.5) and (1.6) gives

$$\frac{d\mathbf{m}}{dt} = \gamma \left[\mathbf{M}_0 \times \mathbf{H}_0 + \mathbf{M}_0 \times \mathbf{h} + \mathbf{m} \times \mathbf{H}_0 + \mathbf{m} \times \mathbf{h} \right]. \tag{1.7}$$

This equation may be simplified further by assuming that $\mathbf{M}_0$ and $\mathbf{H}_0$ are collinear, and therefore have zero cross product.[5] The final term on the right hand side of Eq. (1.7) may also be neglected since it is assumed that $\mathbf{m}$ and $\mathbf{h}$ are small in amplitude compared to the static contributions. We then obtain

$$\frac{d\mathbf{m}}{dt} = \gamma \left[\mathbf{M}_0 \times \mathbf{h} + \mathbf{m} \times \mathbf{H}_0 \right]. \tag{1.8}$$

---

[5]This assumption is based on a negligible anisotropy and a saturated single domain material.

Let us now assume, without loss of generality, that the static fields both point in the $\hat{z}$ direction. Furthermore, let the deviations of the magnetisation in this direction be unchanged to first order, such that $m_z(t) = 0$, and $M_{0z} = M_{\mathrm{sat}}$. Also, supposing that the time-dependence is of the form $e^{-i\omega t}$, then Eq. (1.8) can be written

$$-i\omega \mathbf{m} = \hat{z} \times \left[-\omega_{\mathrm{M}} \mathbf{h} + \omega_0 \mathbf{m} \right], \tag{1.9}$$

where we have defined

$$\omega_{\mathrm{M}} \equiv -\gamma M_{\mathrm{sat}}, \tag{1.10}$$

and

$$\omega_0 \equiv -\gamma H_0. \tag{1.11}$$

We are now in a position to solve Eq. (1.9), which has been linearised with respect to $\mathbf{m}$ and $\mathbf{h}$. Solving for $\mathbf{h}$ gives

$$\begin{pmatrix} h_x \\ h_y \end{pmatrix} = \frac{1}{\omega_{\mathrm{M}}} \begin{pmatrix} \omega_0 & i\omega \\ -i\omega & \omega_0 \end{pmatrix} \begin{pmatrix} m_x \\ m_y \end{pmatrix}. \tag{1.12}$$

The equation can then be inverted to obtain the familiar form relating magnetisation, susceptibility, and magnetic field,

$$\mathbf{m} = \bar{\chi} \cdot \mathbf{h}, \tag{1.13}$$

where $\bar{\chi}$ is the Polder susceptibility tensor and is written as

$$\bar{\chi} = \begin{pmatrix} \chi & -i\kappa \\ i\kappa & \chi \end{pmatrix}, \tag{1.14}$$

and we have defined

$$\chi \equiv \frac{\omega_0 \omega_{\mathrm{M}}}{\omega_0^2 - \omega^2}, \tag{1.15}$$

and

$$\kappa \equiv \frac{\omega \omega_{\mathrm{M}}}{\omega_0^2 - \omega^2}. \tag{1.16}$$

Equation (1.13) describes the change in magnetisation as a function of an applied time-varying small magnetic field. We also note that the components of the Polder susceptibility tensor tend to infinity as $\omega \to \omega_0$. Physically, this infinity represents ferromagnetic resonance, and it may be mathematically avoided if damping mechanisms are included.

1.3.2 Walker's Equation

In the limit of long magnon wavelengths, one can apply the magnetostatic wave approximation.[6] Working within this approximation describes a regime in which spins are dominated by dipolar coupling. The approximation is valid when $k_0(\omega) \ll k(\omega) \ll \pi/a$, where k is the magnon wavenumber, k_0 is the wavenumber of the electromagnetic wave at the same frequency, and a is the spacing between spins.

The magnetostatic equations are

$$\nabla \times \mathbf{h} = 0, \tag{1.17}$$

$$\nabla \cdot \mathbf{b} = 0, \tag{1.18}$$

$$\nabla \times \mathbf{e} = i\omega\mathbf{b}, \tag{1.19}$$

where $\mathbf{h}$, $\mathbf{b}$ and $\mathbf{e}$ represent small amplitude time-varying components of the magnetic field, magnetic flux density, and electric field, respectively. We apply the constitutive relation for a magnetised ferrite[7] to obtain

$$\mathbf{b} = \bar{\mu} \cdot \mathbf{h} \tag{1.20}$$

where

$$\bar{\mu} = \mu_0\big(\bar{\mathbf{I}} + \bar{\chi}\big). \tag{1.21}$$

Using the same susceptibility tensor as in Eq. (1.14), and assuming a magnetic bias field along the $\hat{z}$ direction, then the permeability tensor is

$$\bar{\mu} = \mu_0 \begin{pmatrix} 1+\chi & -i\kappa & 0 \\ i\kappa & 1+\chi & 0 \\ 0 & 0 & 1 \end{pmatrix}. \tag{1.22}$$

We note that since for any analytic function ψ, the vector identity $\nabla \times (\nabla\psi) = 0$ is true, then it must be possible (from Eq. (1.17)) to write

$$\mathbf{h} = \nabla\psi, \tag{1.23}$$

---

[6]For a derivation of this approximation see Ref. [40].

[7]This is the small signal version of the familiar equation, $\mathbf{B} = \mu_0(\mathbf{H} + \mathbf{M})$ that describes the magnetic field in a material.

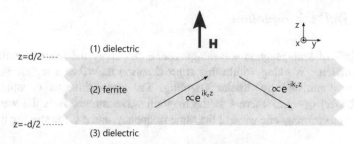

Fig. 1.7 Geometry for forward volume magnetostatic spin waves. The magnetic field is perpendicular to the plane of the film. Spin waves travelling in the film will reflect from the ferrite–dielectric interfaces

where ψ is defined as the magnetostatic scalar potential. Combining this equation with (1.18), and (1.20) we obtain

$$\nabla \cdot \left(\bar{\mu} \cdot \nabla \psi \right) = 0, \tag{1.24}$$

which may then be expanded using Eq. (1.22) to arrive at

$$(1 + \chi)\left[\frac{\partial^2 \psi}{\partial x^2} + \frac{\partial^2 \psi}{\partial y^2} \right] + \frac{\partial^2 \psi}{\partial z^2} = 0. \tag{1.25}$$

This equation is called Walker's equation and is central to calculating the dispersion relations of dipolar spin waves in homogeneous materials. Since this equation was derived using the magnetostatic approximation, the solutions to Walker's equation are referred to as magnetostatic modes.

1.3.3   Forward Volume Modes

As alluded to earlier, the dispersion of dipolar spin waves is highly anisotropic and dependent on field geometry. What follows is a derivation of the dispersion relation relevant to experiments described in later chapters, using Walker's equations and appropriate boundary conditions.[8]

We begin by considering the geometry shown in Fig. 1.7, wherein a ferrite film covers the x–y plane, with a dielectric material above and below, and the biasing magnetic field is in the z direction, parallel with the film normal. The ferrite film has thickness d and is centred on the $z = 0$ plane.

Any spin wave excited in the film will have a wavevector that we describe as

$$\mathbf{k} = \mathbf{k}_t + k_z \hat{\mathbf{z}}, \tag{1.26}$$

---

[8]This treatment follows that offered by Stancil and Prabhakar in *Spin Waves*, Ref. [40].

where

$$\mathbf{k}_t = k_x \hat{\mathbf{x}} + k_y \hat{\mathbf{y}}, \tag{1.27}$$

and it describes the in-plane component of the wavevector. Considering that reflections take place from the upper and lower boundaries, we expect from symmetry that the system will eventually have the same number of waves with $+k_z$ as there are waves with $-k_z$. Given this, we may now make an educated guess at a solution for ψ in the ferrite. Let us suppose it is of the form

$$\psi^{(2)} = \psi_0 \, e^{i\mathbf{k}_t \cdot \mathbf{r}} \left[\frac{e^{ik_z z} + e^{-ik_z z}}{2} \right] \tag{1.28}$$
$$= \psi_0 (k_z z) e^{i\mathbf{k}_t \cdot \mathbf{r}}$$

where ψ_0 is a constant that represents the amplitude of the mode.

Inspecting Eq. (1.25), we note that when $\chi = 0$, as is the case within the dielectric, then Walker's equation simply reduces to Laplace's equation. Assuming that in the dielectric medium, $\psi \propto e^{i\mathbf{k}\cdot\mathbf{r}}$, then applying Walker's equation gives the condition

$$k_{t,d}^2 + k_{z,d}^2 = 0 \tag{1.29}$$
$$\implies k_{z,d} = \pm i k_{t,d}.$$

The subscript d is in reference to the dielectric regions (1) and (3). This suggests a trial solution of

$$\psi_{\text{trial}} = C e^{i\mathbf{k}_{t,d}\cdot\mathbf{r} \pm k_{t,d} z}. \tag{1.30}$$

This solution is at risk of blowing up as the magnitude of z gets large. To prevent this, and to confine the mode to the ferrite, we choose the solutions that go to zero at large distances from the film:

$$\psi^{(1)} = C e^{i\mathbf{k}_{t,d}\cdot\mathbf{r} - k_{t,d} z} \tag{1.31}$$
$$\psi^{(3)} = D e^{i\mathbf{k}_{t,d}\cdot\mathbf{r} + k_{t,d} z}. \tag{1.32}$$

At this point we must apply the boundary conditions of the electromagnetic fields, namely:

• $\mathbf{h}_\parallel$ is continuous across a boundary
• $\mathbf{b}_\perp$ is continuous across a boundary.

Starting with the condition on $\mathbf{h}_\parallel$, we note that from Eq. (1.23),

$$\mathbf{h}_\parallel = \nabla_\parallel \psi, \tag{1.33}$$

where

$$\nabla_\parallel = \hat{\mathbf{x}}\frac{\partial}{\partial x} + \hat{\mathbf{y}}\frac{\partial}{\partial y}. \tag{1.34}$$

The boundary condition therefore becomes

$$\nabla_\parallel \psi^{(1)}(z = d/2) = \nabla_\parallel \psi^{(2)}(z = d/2) \tag{1.35}$$

$$\implies i\mathbf{k}_{\mathrm{t,d}}\psi^{(1)}(z = d/2) = i\mathbf{k}_{\mathrm{t}}\psi^{(2)}(z = d/2), \tag{1.36}$$

and similarly for the $z = -d/2$ boundary. Since our current definitions of $\psi^{(1,2)}$ show that they depend on x and y, the only way for condition Eq. (1.36) to hold everywhere in the plane is for $\mathbf{k}_{\mathrm{t,d}} = \mathbf{k}_{\mathrm{t}}$. Applying this condition to Eq. (1.36) shows that ψ must be continuous, and expansion in terms of Eqs. (1.31) and (1.28) gives

$$Ce^{-k_{\mathrm{t}}d/2} = \psi_0\,(k_z d/2) \tag{1.37}$$

$$De^{-k_{\mathrm{t}}d/2} = \psi_0\,(k_z d/2), \tag{1.38}$$

implying that $C = D$.

Let us now consider the second requirement that $\mathbf{b}_\perp$ must be continuous at the boundaries $z = \pm d/2$. We note from (1.20) and (1.22), that

$$\mathbf{b}_\perp = \mu_0(\nabla\psi)_z, \tag{1.39}$$

which, when applied to the magnetostatic potentials at the boundaries gives

$$k_{\mathrm{t}}Ce^{-k_{\mathrm{t}}d/2} = \psi_0\,k_z(k_z d/2) \tag{1.40}$$

$$-k_{\mathrm{t}}De^{-k_{\mathrm{t}}d/2} = -\psi_0\,k_z(k_z d/2). \tag{1.41}$$

Since we have established that $C = D$, these two equations are identical. We combine them with Eq. (1.37) to obtain

$$\tan(k_z d/2) = \frac{k_{\mathrm{t}}}{k_z}. \tag{1.42}$$

We may obtain another equation relating k_{t} and k_z, by applying Walker's equation to the magnetostatic potential within the ferrite, $\psi^{(2)}$ to find

$$(1 + \chi)k_{\mathrm{t}}^2 + k_z^2 = 0, \tag{1.43}$$

or

$$\frac{k_t}{k_z} = \frac{1}{\sqrt{-(1+\chi)}}. \tag{1.44}$$

Finally, combining Eqs. (1.42) and (1.44), we may eliminate k_z to obtain the relation

$$\tan\left[\frac{k_t d}{2}\sqrt{-(1+\chi)}\right] = \frac{1}{\sqrt{-(1+\chi)}}. \tag{1.45}$$

This is the dispersion relation of dipolar spin waves in a perpendicularly magnetised thin film. This equation relates k_t, the wavenumber within the film, to ω, the angular frequency. The angular frequency does not appear explicitly and is contained with χ (Eq. (1.15)), as is the dependence on the static magnetic field (Eq. (1.11)). Due to the periodicity of the tangent function, there are multiple solutions to Eq. (1.45) which represent different thickness modes, where the nth thickness mode has n zeros throughout the thickness of the film.

An approximation [41] of the first mode is given by

$$\omega = \sqrt{\omega_0\left[\omega_0 + \omega_M\left(1 - \frac{1 - e^{-k_t d}}{k_t d}\right)\right]}. \tag{1.46}$$

Waves of this sort are referred to as 'forward volume magnetostatic spin waves' (FVMSW). The dispersion curves of the first 3 thickness modes are plotted in Fig. 1.8.

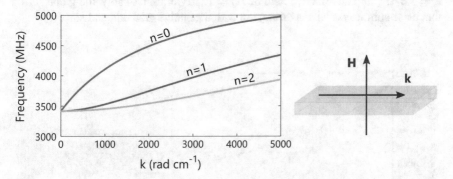

Fig. 1.8 Dispersion curve and field geometry for forward volume magnetostatic spin waves (FVMSW). The magnetic field is parallel to the film normal and perpendicular to the wavevector. The branches for the first three thickness modes are plotted for $M_{sat} = 140\,\text{kAm}^{-1}$, a film thickness $d = 7.8\,\mu\text{m}$, and an applied field of $H_0 = 3000\,\text{Oe}$

1.3.4   Backward Volume Modes

Having treated the case when the applied magnetic field is normal to the film, it was found that FVMSW propagate isotropically within the plane. This result is not unexpected due to the symmetry of the system. Once the symmetry is broken and the field is in plane, the magnetostatic mode behaviour is much richer. Indeed, magnons propagating parallel to the applied magnetic field behave differently from those propagating perpendicular. For the sake of brevity, a full derivation of these modes is not presented, only the key results.

Consider the same geometry as in Fig. 1.7, but with the field applied in the y direction. After a similar analysis as above, the dispersion relation for spin waves propagating parallel with the field is

$$\tan\left[\frac{k_y d}{2\sqrt{-(1+\chi)}}\right] = \sqrt{-(1+\chi)}. \tag{1.47}$$

Again, the periodicity of the tangent implies multiple thickness modes. The lowest order mode however may be approximated [41] to a form that makes explicit the relationship between ω and k:

$$\omega = \sqrt{\omega_0\left[\omega_0 + \omega_M\left(\frac{1 - e^{-k_y d}}{k_y d}\right)\right]}. \tag{1.48}$$

This result is plotted in Fig 1.9 for parameters $M_{\text{sat}} = 140\,\text{kAm}^{-1}$, a film thickness $d = 7.8\,\mu\text{m}$, and an effective field of $H_0 = 1220\,\text{Oe}$. Remarkably, the graph shows that these spin waves have a dispersion with a negative gradient, and since

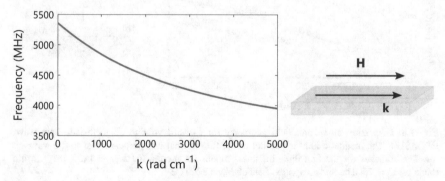

Fig. 1.9 Dispersion curve and field geometry for backward volume magnetostatic spin waves (BVMSW). The static field lies in the plane of the film and is parallel to the wavevector. The first thickness mode is plotted for $M_{\text{sat}} = 140\,\text{kAm}^{-1}$, a film thickness $d = 7.8\,\mu\text{m}$, and an effective field of $H_0 = 1220\,\text{Oe}$. The negative gradient corresponds to a negative group velocity

$$v_g = \frac{\partial \omega}{\partial k},\tag{1.49}$$

this implies a negative group velocity, v_g. Meanwhile, the phase velocity,

$$v_p = \frac{\omega}{k},\tag{1.50}$$

is positive. Thus, the phase velocity and group velocity are antiparallel. Because of the opposing velocities, the waves are therefore referred to as backwards volume magnetostatic spin waves (BVMSW).

1.3.5 Magnetostatic Surface Waves

The remaining geometry to consider is with the field in plane, and the magnons travelling in the x direction, perpendicular to the field. As with Sect. 1.3.4, consider the field pointing in the y direction. The dispersion relation for these waves is

$$\omega = \sqrt{\omega_0(\omega_0 + \omega_M) + \frac{\omega_M^2}{4}(1 - e^{-k_x d})}.\tag{1.51}$$

Unlike the similar-form expressions for the volume modes, this equation is exact and not an approximation of the first solution of a periodic function.

These waves have several important characteristics. There is only a *single* mode, defined by Eq. (1.51), rather than a series of modes that vary differently throughout the film thickness. Indeed, in contrast to the volume modes discussed above, the amplitude of the mode is not distributed sinusoidally throughout the film, but decays

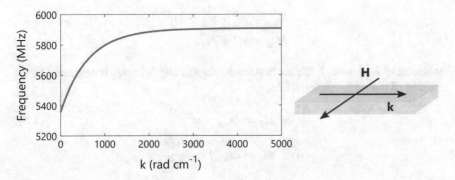

Fig. 1.10 Dispersion curve and field geometry for magnetostatic surface spin waves (MSSW). The static field, the film normal and the wavevectors are mutually orthogonal. There is only a single mode described by Eq. (1.51) which is plotted for $M_{\text{sat}} = 140\,\text{kAm}^{-1}$, a film thickness $d = 7.8\,\mu\text{m}$, and an effective field of $H_0 = 1220\,\text{Oe}$

exponentially from the surfaces. Thus, these waves are named magnetostatic surface spin waves (MSSW).

Another distinguishing feature of MSSW is the phenomenon of field displacement non-reciprocity. The direction of $\mathbf{k}$ of the surface mode depends on $\mathbf{H}$ and the surface normal; if the field direction is reversed, the surface mode will shift from one side of the film to the other (Fig. 1.10).

1.4 Nonlinear Behaviour

The inherent nonlinearity of spin waves has lead to much interesting physical phenomena, including auto-oscillations, chaotic dynamics, strange attractors, spin wave envelope solitons, and time-domain fractals [42, 43]. These effects may be generated at much lower wave-power than in an optical setting, due to the relatively large nonlinearity of spin waves. Furthermore, the slow propagation speeds (compared with light waves) mean that time-domain measurements may be readily performed using fast, modern electronics.

It is well established that in materials possessing spontaneous long range magnetic order, such as ferromagnets, the exchange interaction between spins is strong enough such that the total magnitude of the magnetisation vector $\mathbf{M}$ is constant [44] and is

$$|\mathbf{M}| = M_{\text{sat}} = M_z \qquad (1.52)$$

for a film magnetised in the z direction. If the film experiences a transverse magnetisation $\mathbf{m}$ in the plane of the film, the magnetisation along the z axis will be reduced. The relationship between the saturation magnetisation and the components of the magnetisation vector then becomes

$$M_{\text{sat}}^2 = m_x^2 + m_y^2 + M_z^2 \qquad (1.53)$$
$$M_{\text{sat}}^2 = m^2 + M_z^2,$$

where $\mathbf{m} = m_x \hat{\mathbf{x}} + m_y \hat{\mathbf{y}}$. We see that the demagnetizing field, M_z is decreased by the canting of the total magnetisation:

$$M_z^2 = M_{\text{sat}}^2 - m^2 \qquad (1.54)$$

$$\implies M_z = M_{\text{sat}} \sqrt{1 - \frac{m^2}{M_{\text{sat}}^2}}$$

$$\approx M_{\text{sat}} \left(1 - \frac{m^2}{2M_{\text{sat}}^2}\right)$$

$$= M_{\text{sat}} \left(1 - |\Psi|^2\right),$$

where we have defined $\Psi^2 = m^2/2M_{sat}^2$ as the dimensionless amplitude of transverse magnetisation [45]. As the spin wave amplitude increases, the demagnetising field decreases, changing the value of ω_M and ω_H, thus also the frequency of $\omega(k)$. This is the origin of nonlinearity in magnetodynamical systems.

1.4.1 Lighthill Criteria

A nonlinear wave phenomenon common to many areas of physics [46] is modulational instability. When the amplitude of a nonlinear wave—the carrier—is above a certain threshold, it can undergo a process wherein the carrier at a frequency ω, is spontaneously modulated by side bands at frequencies $\omega \pm \Omega$.

A condition for modulational instability of a nonlinear stationary wave was first reached by Lighthill [47] in the 1960s. It was shown that the square of the speed of a wave with amplitude Ψ_0 was equal to

$$c_0^2 = \Psi_0^2 \frac{1}{v_{g0}'} \left(\frac{\partial \omega}{\partial(\Psi^2)} \right)_0. \tag{1.55}$$

where v_{g0}' is the derivative of the group velocity with respect to wavenumber at the amplitude Ψ_0. From that result it followed that if $c_0^2 < 0$, that is,

$$\frac{1}{v_{g0}'} \left(\frac{\partial \omega}{\partial(\Psi^2)} \right)_0 < 0, \tag{1.56}$$

then the wave becomes unstable and modulation occurs [48]. This condition has become known as the Lighthill criteria, and is more commonly expressed as

$$ND < 0, \tag{1.57}$$

where

$$N = \left(\frac{\partial \omega}{\partial(\Psi^2)} \right)_0, \tag{1.58}$$

and

$$D = v_{g0}' = \left(\frac{\partial^2 \omega}{\partial k^2} \right)_0. \tag{1.59}$$

The quantities N and D are referred to as the nonlinearity and dispersion coefficients, respectively, and arise from a nonlinear Schrodinger (NLS) equation formalism [49].

It may be noted by inspection of the FVMSW dispersion shown in Fig. 1.8, that $D < 0$ for the n=0 mode for all k. Furthermore, in the limit of small k in FVMSW, it may be shown using Eq. (1.46) that N is equal to ω_M [50], which is always positive. The Lighthill criterion is therefore always satisfied for FVMSWs meaning that modulational instability is possible.

References

1. Waldrop MM (2016) The chips are down for Moore's law. Nature 530:144–147
2. Kalin I, Ayduz S, Dagli C (2013) The Oxford encyclopedia of philosophy, science, and technology in Islam. Oxford University Press, Oxford
3. Binasch G, Grünberg P, Saurenbach F, Zinn W (1989) Enhanced magnetoresistance in layered magnetic structures with antiferromagnetic interlayer exchange. Phys. Rev. B 39:4828–4830
4. Baibich MN et al (1988) Giant magnetoresistance of (001)Fe/(001)Cr magnetic superlattices. Phys. Rev. Lett. 61:2472–2475
5. Moore GE (1965) Cramming more components onto integrated circuits. Electron, Mag
6. Dennard R, Gaensslen F, Rideout V, Bassous E, LeBlanc A (1974) Design of ion-implanted MOSFET's with very small physical dimensions. IEEE J Solid-State Circuits 9:256–268
7. Yangyuan W, Ruqi H, Xiaoyan L, Jinfeng K (1998) The challenges for physical limitations in Si microelectronics. In: 1998 5th International Conference on Solid-State and Integrated Circuit Technology. Proceedings (Cat. No.98EX105), pp 25–30
8. Frank DJ (2002) Power-constrained CMOS scaling limits. IBM J Res Dev 46:235–244
9. Haron NZ, Hamdioui S (2008) Why is CMOS scaling coming to an END? In: 2008 3rd International Design and Test Workshop. pp 98–103
10. Shankland S (2018) iPhone XS A-12 Bionic chip is industry-first 7nm CPU-CNET. https://www.cnet.com/news/iphone-xs-a12-bionic-chip-is-industry-first-7nm-cpu/
11. Bohr M (2007) A 30 year retrospective on Dennard's MOSFET scaling paper. IEEE Solid-State Circuits Newsl 12:11–13
12. Nikonov DE, Young IA (2013) Overview of beyond-CMOS devices and a uniform methodology for their benchmarking. Proc IEEE 101:2498–2533
13. Chumak AV, Vasyuchka V, Serga AA, Hillebrands B (2015) Magnon spintronics. Nat Phys 11:453–461
14. Dhillon SS et al (2017) The 2017 terahertz science and technology roadmap. J Phys D Appl Phys 50:043001
15. Zografos O et al (2016) Design and benchmarking of hybrid CMOS-spin wave device circuits compared to 10 nm CMOS. IEEE-NANO 2015 - 15th International Conference on Nanotechnology. pp 686–689
16. Kostylev MP, Serga AA, Schneider T, Leven B, Hillebrands B (2005) Spinwave logical gates. Appl Phys Lett 87:1–3
17. Schneider T et al (2018) Realization of spin-wave logic gates. Appl Phys Lett 92:0-3. arXiv: 0711.4720
18. Chumak AV, Serga A, Hillebrands B (2014) Magnon transistor for allmagnon data processing. Nat Commun 5:4700
19. Gulyaev YV et al (2003) Ferromagnetic films with magnon bandgap periodic structures: magnon crystals. J Exp Theor Phys Lett 77:567–570
20. Cornelissen LJ, Liu J, van Wees BJ, Duine RA (2018) Spin-current- controlled modulation of the magnon spin conductance in a three-terminal magnon transistor. Phys Rev Lett 120:097702
21. Wu H et al (2018) Magnon valve effect between two magnetic insulators. Phys Rev Lett 120:097205
22. Tock CJ, Gregg JF (2019) Phase modulation and amplitude modulation interconversion for magnonic circuits. Phys Rev Appl 11:044065
23. Balynsky M et al (2017) Magnonic interferometric switch for multi-valued logic circuits. J Appl Phys 121:024504

24. Feitelson DG (1988) Optical computing: a survey for computer scientists. MIT Press, Cambridge
25. Csaba G, Papp á, Porod W (2017) Perspectives of using spin waves for computing and signal processing. Phys Lett A 1:1–6
26. Toedt J-N, Mundkowski M, Heitmann D, Mendach S, Hansen W (2016) Design and construction of a spin-wave lens. Sci Rep 6:33169
27. Jeon D, Seok M, Chakrabarti C, Blaauw D, Sylvester D (2012) A super-pipelined energy efficient subthreshold 240 MS/s FFT core in 65 nm CMOS. IEEE J Solid-State Circuits 47:23–34
28. Papp á, Porod W, Csurgay áI, Csaba G (2017) Nanoscale spectrum analyzer based on spin-wave interference. Sci Rep 7:9245 (2017)
29. Csaba G, Papp A, Porod W (2014) Spin-wave based realization of optical computing primitives. J Appl Phys 115:17C741
30. Chumak AV et al (2010) All-linear time reversal by a dynamic artificial crystal. Nat Commun 1:141
31. Krawczyk M, Grundler D (2014) Review and prospects of magnonic crystals and devices with reprogrammable band structure. J Phys Condens Matter 26:123202
32. Chumak AV, Serga AA, Hillebrands B (2017) Magnonic crystals for data processing. J Phys D Appl Phys 50:244001
33. Karenowska AD et al (2012) Oscillatory energy exchange between waves coupled by a dynamic artificial crystal. Phys Rev Lett 108:015505. arXiv:1106.3722
34. Kittel C (2004) Introduction to solid state physics. Wiley, New York
35. Ashcroft N, Mermin N (1976) Solid state physics. Saunders College
36. Mruczkiewicz M et al (2017) Spin-wave nonreciprocity and magnonic band structure in a thin permalloy film induced by dynamical coupling with an array of Ni stripes. Phys Rev B 96:104411
37. Chen J et al (2019) Excitation of unidirectional exchange spin waves by a nanoscale magnetic grating. arXiv: 1903.00638
38. Karenowska AD, Chumak AV, Serga AA, Hillebrands B (2016) Handbook of spintronics. Springer Netherlands, pp 1505–1549
39. Cherepanov V, Kolokolov I, L'vov V (1993) The saga of YIG: spectra, thermodynamics, interaction and relaxation of magnons in a complex magnet. Phys Rep 229:81–144
40. Prabhakar A, Stancil DD (2009) Spin waves. Springer, Boston
41. Kalinikos B (1980) Excitation of propagating spin waves in ferromagnetic films. IEE Proc H Microwaves, Opt Antennas 127:4
42. Slavin AN, Demokritov SO, Hillebrands B (2001) Spin dynamics in confined magnetic structures I. Springer, Berlin, pp 35–66
43. Richardson D, Kalinikos BA, Carr LD, Wu M (2018) Spontaneous exact spin-wave fractals in magnonic crystals. Phys Rev Lett 121:107204
44. Lax B, Button KJ (1962) Microwave ferrites and ferrimagnetics. McGraw-Hill, New York
45. Kalinikos BA, Kovshikov NG, Slavin AN (1991) Envelope solitons of highly dispersive and low dispersive spin waves in magnetic films. J Appl Phys 69:5712–5717
46. Zakharov VE, Ostrovsky LA (2009) Modulation instability: the beginning. Phys D Nonlinear Phenom 238:540–548
47. Lighthill MJ (1965) Contributions to the theory of waves in non-linear dispersive systems. IMA J Appl Math 1:269–306
48. Karpman VI (1975) Non-linear waves in dispersive media. Pergamon, Oxford
49. Slavin A, Rojdestvenski I (1994) "Bright" and "dark" spin wave envelope solitons in magnetic films. IEEE Trans Magn 30:37–45
50. Tsankov MA, Chen M, Patton CE (1994) Forward volume wave microwave envelope solitons in yttrium iron garnet films: propagation, decay, and collision. J Appl Phys 76:4274–4289

Chapter 2
Magnonic Phase Conjugation Theory

In everyday experience time always moves forward. The situation is qualitatively different, however, in the case of wave motion: light waves can be "time-reversed" and made to retrace their trajectories

Boris Zel'dovich Optical Phase Conjugation [1]

Building on the general introduction of magnon computing and the motivations thereof presented in Chap. 1, in this chapter we concentrate on a specific operation that may be exploited, namely phase conjugation. We briefly introduce the concept and derive the main result for the well-established optical case. Inspired by this, we present for the first time, a result central to this thesis: an original theoretical treatment of magnon phase conjugation arising from four-wave mixing.

2.1 Phase Conjugation: What Is It and What Does It Do?

Phase conjugation is a technique that exploits a nonlinear system to exactly reverse the direction of travel and phase factor for each and every plane wave comprising an arbitrary wavefront. This process results in a novel type of mirror; the incoming wave is reflected, but unlike a conventional mirror, the phase conjugate reflection retraces the original path precisely. A simple illustration of this may be seen in Fig. 2.1 where both types of reflection are compared. The phase conjugate mirror (PCM) performs an operation on two different parameters of the incident plane wave: the $\mathbf{k}$-vector and the phase. The action of a traditional mirror on a wave can be described as the reversal of the $\mathbf{k}$-vector component that is normal to the plane of the mirror. For example, a mirror in the x-y plane will reverse the k_z component of the incoming wave. Explicitly, we can describe this process as

A. Inglis, *Investigating a Phase Conjugate Mirror for Magnon-Based Computing*, Springer Theses, https://doi.org/10.1007/978-3-030-49745-3_2

traditional mirror   phase conjugate mirror

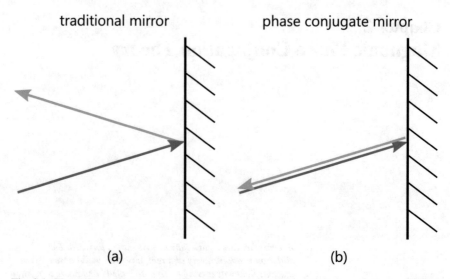

(a)   (b)

Fig. 2.1 Illustration comparing the difference between a traditional mirror and phase conjugate mirror (PCM). The incoming rays (blue) are reflected (red) from a traditional mirror (**a**) in the expected way, while the rays reflected from the PCM retrace exactly the same path as the incoming ray (color online)

$$\mathbf{k}_{\text{in}} = k_x\hat{\mathbf{x}} + k_y\hat{\mathbf{y}} + k_z\hat{\mathbf{z}}$$
$$\mathbf{k}_{\text{out}} = k_x\hat{\mathbf{x}} + k_y\hat{\mathbf{y}} - k_z\hat{\mathbf{z}}, \tag{2.1}$$

which is illustrated in Fig. 2.1a. On the other hand, if we wish to consider the action of a PCM in an analogous way, we must imagine a mirror that has its plane *defined by the **k**-vector of the incoming wave*, such that any incoming wave is entirely reversed. That is,

$$\mathbf{k}_{\text{in}} = -\mathbf{k}_{\text{out}} \tag{2.2}$$

for any plane wave incident on a PCM. Since this is true for plane waves it is also the case for an arbitrary wavefront composed of any number of plane waves. Figure 2.1b illustrates the action of the PCM on an incoming wave. The physical plane of the PCM is fixed, as it would be in an experiment, but the *action on the wavevector* appears as if it depends on the incoming wave.

In addition to the novel operation on the momentum of the incident wave, there is also the transformation of the phase. A plane wave, F, travelling in the positive-z direction with complex amplitude, A, can be written as

$$F_{\text{in}} = Ae^{i(kz-\omega t)}$$
$$= |A|e^{i\phi}e^{i(kz-\omega t)}. \tag{2.3}$$

In the case where F were to be reflected off a PCM, not only would k be negated, but so too would the phase, ϕ, resulting in the following:

$$F_{\text{out}} = A^* e^{i(-kz-\omega t)}$$
$$= |A| e^{-i\phi} e^{i(-kz-\omega t)}. \tag{2.4}$$

We see that upon reflection from the PCM, the wave undergoes a transformation

$$(\phi, k, \omega) \rightarrow (-\phi, -k, \omega). \tag{2.5}$$

The reflected wave now has a relative sign difference between ω and the other two parameters, ϕ and k. For this reason, the effect is occasionally referred to as 'time reversal'.

2.2 Optical Phase Conjugation

Optical phase conjugation can be achieved by exploiting a number of different nonlinear optical effects. The variation of approaches makes the subject useful for demonstrating similarities between certain nonlinear processes, but is also contextualised by the fact that this field was developed independently by various groups simultaneously. The complexity with which the study of optical phase conjugation evolved is epitomised in a historical overview when Robert Fisher writes [2]: 'Searching for a progenitor of optical phase conjugation appears to be as futile as trying to find the father of a new litter of alley cats.'

Historically, there have been three main approaches to optical phase conjugation. These disciplines were transient holography [3], stimulated inelastic processes [4], and elastic parametric scattering processes in nonlinear media [5]. Of these, we shall narrow our focus on an aspect of the latter approach: optical phase conjugation by four-wave mixing.

2.2.1 Aberration Correction: An Interesting Digression

Before delving into the specifics of four-wave mixing, a general yet remarkable property of phase conjugate waves must be briefly explored. The distortion-undoing nature of phase conjugate reflections give rise to some very useful applications [6–8] of this technique. What follows is a mathematical proof[1] of this process. The treatment, though performed for electromagnetic waves, is without loss of generality.

We consider a plane wave, $\tilde{E}_s (\mathbf{r}, t)$ propagating in the $+z$ direction, taking the form:

---

[1]For a similar treatment see Boyd, *Nonlinear Optics* [9].

$$\tilde{E}_s\left(\mathbf{r}, t\right) = E_s\left(\mathbf{r}\right) e^{-i\omega t} + \text{c.c.}$$
$$= A_s(\mathbf{r})e^{(kz-i\omega t)} + \text{c.c.}, \tag{2.6}$$

where A_s is the complex amplitude of the plane wave. Let us introduce a distorting medium of non-uniform refractive index $n(\mathbf{r}) = \sqrt{\epsilon(\mathbf{r})}$ with the assumption that the scale over which $\epsilon(\mathbf{r})$ varies is much larger than $2\pi/k$. The electric field must obey the wave equation which may be written as [10]

$$\left[\nabla_T^2 + \frac{\partial^2}{\partial z^2}\right]\tilde{E}_s - \frac{\epsilon(\mathbf{r})}{c^2}\frac{\partial^2 \tilde{E}_s}{\partial t^2} = 0, \tag{2.7}$$

where the Laplacian operator has been expanded using the transverse Laplacian $\nabla_T^2 = \frac{\partial^2}{\partial x^2} + \frac{\partial^2}{\partial y^2}$. Combining Eqs. (2.6) and (2.7), we obtain the relation

$$\nabla_T^2 A_s + \left[\frac{\omega^2 \epsilon(\mathbf{r})}{c^2} - k^2\right] A_s + 2ik\frac{dA_s}{dz} = 0. \tag{2.8}$$

In arriving at Eq. (2.8) we have applied the slowly varying approximation that allows us to drop the $\frac{\partial^2 A_s}{\partial z^2}$ term. Note however that since this equation is mathematically valid, so too must its complex conjugate be; that is,

$$\nabla_T^2 A_s^* + \left[\frac{\omega^2 \epsilon(\mathbf{r})}{c^2} - k^2\right] A_s^* - 2ik\frac{\partial A_s^*}{\partial z} = 0. \tag{2.9}$$

We now observe that the perfectly legitimate Eq. (2.9) describes the wave

$$\tilde{E}_c\left(\mathbf{r}, t\right) = A_c(\mathbf{r})e^{(-kz-i\omega t)} + \text{c.c.}, \tag{2.10}$$

under the condition that

$$A_c(\mathbf{r}) = A_s^*(\mathbf{r}). \tag{2.11}$$

The wave described by Eq. (2.10) is travelling in the $-z$ direction with a complex amplitude A_c that is *everywhere* the complex conjugate of A_s. This implies that given a plane wave with amplitude A_s passing through some medium that distorts the wavefronts, if we can create a phase conjugate reflection with complex amplitude A_c then it will propagate back though the the aberrating medium and recover the original plane wave. This process is illustrated in Fig. 2.2, where wavefronts containing the information of an image pass through a distorting medium such as a turbulent atmosphere, or a poorly constructed optic. In Fig. 2.2a the waves are reflected from a traditional mirror, whereas Fig. 2.2b depicts reflection from a phase conjugate mirror. We see that after reflection, when the wavefronts pass through the material of nonuniform refractive index for a second time, the image is distorted for a second time in (a), while in (b) the distortion has been undone. This usual property is another reason that the phenomenon is sometimes called 'time-reversal'.

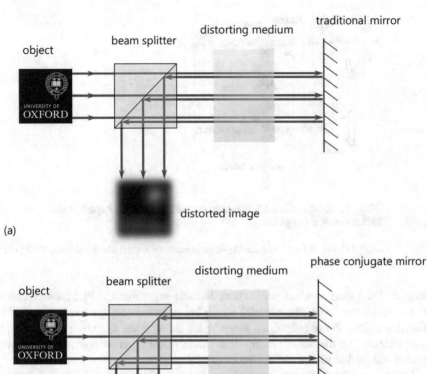

Fig. 2.2 Sketch of an experiment highlighting the ability of a phase conjugate mirror (PCM) to correct aberrations of a wavefront. In both diagrams, wavefronts containing information about the original object are passed through a distorting medium before being reflected and passing through the medium again. In **a** the waves are reflected from a traditional mirror and doubly-distorted, in contrast to **b** where waves are reflected from a PCM and distortion is undone, allowing an image to be recovered

2.2.2 *Optical Phase Conjugation via Four-Wave Mixing*

Having seen an example of what one can do with a phase conjugate mirror, the question still remains as to how to create such a device. We shall now explore a method of creating an optical PCM that will serve as an inspiration for the magnetic case. The original design for an optical PCM was first published by Hellwarth in January 1977 [11]. Six months later, Bloom and Bjorklund reported their observation of such an effect [12] using the experiment shown in Fig. 2.3 which is the original

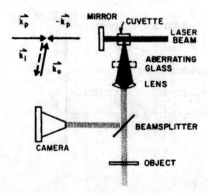

FIG. 1. Schematic of backward–wave generation experiment
and *k*-vector diagram.

Fig. 2.3 Original sketch of four-wave mixing experiment from Bloom and Bjorklund, 1977 [12]

diagram from their seminal publication. In their experiment, a PCM was created
in a nonlinear medium (the cuvette) which led to phase conjugate reflections of a
distorted signal. These reflections passed back through the distorting medium and
were observed by the camera as un-distorted. In this section we see how four-wave
mixing can lead to such a phenomenon.

Let us consider a lossless, isotropic nonlinear medium with third-order susceptibility, $\chi^{(3)}$. Furthermore, let it be illuminated by two counter-propagating pump
waves E_1 and E_2 and a weaker probe wave E_p all of which are degenerate, with an
angular frequency of ω. As a result of the nonlinear coupling of these waves, a fourth
wave E_c appears which is the phase conjugate of E_p. Figure 2.4 shows an example
geometry of such an experiment with the counter-propagating pump waves perpendicular to the probe and conjugate waves which are travelling in the $\pm z$ direction.
For the sake of clarity, we shall treat all waves as plane waves of the form in Eq. (2.6),
with the same assumption that $A_i(\mathbf{r})$ is slowly varying, for all i.

The polarization of the nonlinear medium resulting from the presence of these
electric fields is given by

$$P = 3\chi^{(3)} E^2 E^*, \tag{2.12}$$

where $E = E_1 + E_2 + E_c + E_p$. By choosing this combination of E and E^* we guar-
antee that the resulting polarization will also oscillate at ω and that $\chi^{(3)} = \chi^{(3)}(\omega =
\omega + \omega - \omega)$. Expanding this product results in many cross terms, from which we pay
particular heed to those with a spatial dependence that goes like

$$e^{i\mathbf{k}_j \cdot \mathbf{r}} \quad \text{where} \quad j = 1, 2, p, c \tag{2.13}$$

since terms with these spatial dependencies are phase-matched source terms for one
of the four waves of interest. Collecting the terms with these spatial dependencies,

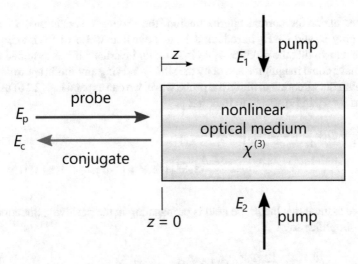

Fig. 2.4 Illustration of a phase conjugate mirror caused by four-wave mixing. The two counter-propagating pumps in a nonlinear medium couple with the incoming probe wave to create a fourth wave that is the phase conjugate of the probe wave

the polarizations associated with them are:

$$P_1 = 3\chi^{(3)} \left[E_1^2 E_1^* + 2E_1 E_2 E_2^* + 2E_1 E_p E_p^* + 2E_1 E_c E_c^* + 2E_p E_c E_2^* \right],$$
$$P_2 = 3\chi^{(3)} \left[E_2^2 E_2^* + 2E_2 E_1 E_1^* + 2E_2 E_p E_p^* + 2E_2 E_c E_c^* + 2E_p E_c E_1^* \right],$$
$$P_p = 3\chi^{(3)} \left[E_p^2 E_p^* + 2E_p E_1 E_1^* + 2E_p E_2 E_2^* + 2E_p E_c E_c^* + 2E_1 E_2 E_c^* \right],$$
$$P_c = 3\chi^{(3)} \left[E_c^2 E_c^* + 2E_c E_1 E_1^* + 2E_c E_2 E_2^* + 2E_c E_p E_p^* + 2E_1 E_2 E_p^* \right]. \quad (2.14)$$

From these polarizations we recall that the pump waves are much stronger than the other waves and so neglect terms with fewer than two factors of pump-field amplitude. We are left with polarizations of the form:

$$P_1 = 3\chi^{(3)} \left[E_1^2 E_1^* + 2E_1 E_2 E_2^* \right],$$
$$P_2 = 3\chi^{(3)} \left[E_2^2 E_2^* + 2E_2 E_1 E_1^* \right],$$
$$P_p = 3\chi^{(3)} \left[2E_p E_1 E_1^* + 2E_p E_2 E_2^* + 2E_1 E_2 E_c^* \right],$$
$$P_c = 3\chi^{(3)} \left[2E_c E_1 E_1^* + 2E_c E_2 E_2^* + 2E_1 E_2 E_p^* \right]. \quad (2.15)$$

We can now apply these equations to the wave equation that includes a polarization driving term [10]:

$$\nabla^2 \tilde{E}_i - \frac{\epsilon}{c^2} \frac{\partial^2 \tilde{E}_i}{\partial t^2} = \frac{4\pi}{c^2} \frac{\partial^2 \tilde{P}_i}{\partial t^2}. \quad (2.16)$$

Note that under the current approximation, the driving polarizations, P_1 and P_2 depend only E_1 and E_2. By introducing the first two lines of Eq. (2.15) into Eq. (2.16) it can be shown that the product $A_1 A_2$ is spatially invariant[2] if it is assumed that A_1 and A_2 have equal magnitude and obey the slowly-varying amplitude approximation.

Turning our attention towards the probe wave, we can expand Eq. (2.16) using the equation for P_p in Eq. (2.15) to find that

$$\left[\left(-k_p^2 + 2ik_p \frac{d}{dz} + \frac{\omega^2 \epsilon}{c^2} \right) A_p \right] e^{i(k_p z - \omega t)}$$

$$= \frac{4\pi\omega^2}{c^2} 3\chi^{(3)} \left[\left(|A_1|^2 + |A_2|^2 \right) A_p + A_1 A_2 A_c^* \right] e^{i(k_p z - \omega t)}, \tag{2.17}$$

where we assume that the probe field is propagating in the positive-z direction. This may be simplified to

$$\frac{dA_p}{dz} = \frac{12\pi i \omega^2}{nc} \chi^{(3)} \left[\left(|A_1|^2 + |A_2|^2 \right) A_p + A_1 A_2 A_c^* \right]. \tag{2.18}$$

Similarly, we can follow the same steps for the polarization P_c to obtain

$$\frac{dA_c}{dz} = -\frac{12\pi i \omega^2}{nc} \chi^{(3)} \left[\left(|A_1|^2 + |A_2|^2 \right) A_c + A_1 A_2 A_p^* \right]. \tag{2.19}$$

These equations describe the coupled amplitudes of the fields E_p and E_c and their spatial dependence. We can write these more conveniently as

$$\frac{dA_p}{dz} = i\kappa_p A_p + i\kappa A_c^*, \tag{2.20a}$$

$$\frac{dA_c}{dz} = -i\kappa_p A_c + i\kappa A_p^*, \tag{2.20b}$$

where constants κ and κ_p describe coupling strength and are defined as

$$\kappa_p = \frac{12\pi i \omega^2}{nc} \chi^{(3)} \left(|A_1|^2 + |A_2|^2 \right), \tag{2.21a}$$

$$\kappa = \frac{12\pi i \omega^2}{nc} \chi^{(3)} A_1 A_2. \tag{2.21b}$$

Equations (2.20) can be further simplified by the change of variables:

$$A_p = A_p' e^{i\kappa_p z} \tag{2.22a}$$

$$A_c = A_c' e^{-i\kappa_p z}, \tag{2.22b}$$

---

[2]The proof for this is fairly straightforward and has been omitted since it follows the same method used to solve Eq. (2.16) for the more interesting waves E_p and E_c. A complete treatment may be found on pages (29–32) in *Optical Phase Conjugation* by Fisher.

where we note that the original variables coincide with the primed version at the plane $z = 0$. The new variables simply represent a change in phase and do not affect the energy exchange between E_p and E_c. Plugging these equations into Eq. (2.20a), we get

$$i\kappa_p A_p' e^{i\kappa_p z} + \frac{dA_p'}{dz} e^{i\kappa_p z} = i\kappa_p A_p' e^{i\kappa_p z} + i\kappa A_c'^* e^{i\kappa_p z}, \tag{2.23}$$

which, by inspection becomes

$$\frac{dA_p'}{dz} = i\kappa A_c'^*, \tag{2.24a}$$

and a similar treatment of Eq. (2.20b) returns

$$\frac{dA_c'}{dz} = -i\kappa A_p'^*. \tag{2.24b}$$

This set of equations first derived in 1977 by Yariv and Pepper [13] shows how degenerate four-wave mixing results in phase conjugation. Equations (2.24) state that the driving amplitude of the generated wave with amplitude A_c' is proportional to the complex conjugate of the probe wave, $A_p'^*$. These simple and seminal equations are a starting point for much of the optical phase conjugation formalism and can be solved for a number of different physical situations, and it is with this result that we move towards a magnetic system.

2.3 Magnonic Phase Conjugation

In comparison to the optical analogue, phase conjugation in magnons remains largely unexplored. The effect was first experimentally demonstrated in 1998 by Gordon et al. [14], but may have been unwittingly excited by Melkov et al. in a similar experiment almost 10 years prior [15]. These works, and the many that followed [16–21] focus exclusively on phase conjugation via three-wave mixing, a second-order effect. The phase conjugate signal is generated from a parametric pumping [22] process whereby a high-power pump decays into a forward and a backward travelling magnon. If this process is stimulated, then these magnons can be interpreted as the amplification of the stimulant spin wave (forward-travelling) and the generation of its phase conjugate reflection (backward-travelling). Aside from focussing only on parametric-pumping, these experiments were also exclusively in the BVMSW regime, in contrast to the treatment that follows.

Having familiarised ourselves with the origin of optical phase conjugation from four-wave mixing, we shall now examine four-wave mixing in a magnetic system. In this section, the physical concepts invoked by the above derivations will be given priority over the strictest rigour of a full mathematical treatment, which remains to be established by the magnonic community. We shall approach the problem of wave-mixing using a perturbation expansion as first outlined in 2012 by Marsh and Camley [23] in their analysis of two-wave mixing. Our treatment is original however, and shows explicitly for the first time the origin of a phase conjugate magnon resulting from four-wave mixing.

2.3.1 Assumptions and Approximations

By transitioning our model from the optical to magnetic regime, we must also update the assumptions that underlie the theory. A number of approximations are common to both treatments. Explicitly, all waves are monochromatic plane waves from a coherent source, propagating in a lossless medium. We also maintain the condition that the pump amplitudes are large compared to the probe amplitude, and that each system is operating in the steady state. Specific to the magnetic case however, is the assumption that the spin wave medium is an infinite plane with magnetisation fully saturated out of plane.

2.3.2 The Landau–Lifschitz Equation

We begin by considering a lossless magnetic thin-film in the x-y plane, with an external magnetic field applied normal to the plane of the film, such that FVMSWs are excited (Fig. 2.5).

We can describe this system using the Landau Lifschitz equation. Namely,

$$\frac{d\mathbf{M}}{dt} = \gamma \mathbf{M} \times \mathbf{H}_{\text{eff}}, \tag{2.25}$$

where $\mathbf{H}_{\text{eff}}$ is the effective field and is defined by the sum

$$\mathbf{H}_{\text{eff}} = H_{\text{ext}}\hat{z} + h_y\hat{y} + h_z\hat{z} - N_x m_x\hat{x} - N_y m_y\hat{y} - N_z M_{\text{sat}}\hat{z}. \tag{2.26}$$

The effective field comprises a large externally applied field H_{ext}, an oscillating magnetic field $h_{y,z}$, and usual contributions from the demagnetisation, N_i that are coupled to the magnetisation. It has been assumed that transverse magnetisations, $m_{x,y}$ are small enough and that the applied static magnetic field is large enough such that the condition $m_z = M_{\text{sat}}$ is valid. Physically, this corresponds to magnons with a small precession angle that allow us to work in the linear regime [24] where

$$\mathbf{M} = (m_x, m_y, M_{\text{sat}}). \tag{2.27}$$

We now introduce Eqs. (2.26) and (2.27) to (2.25) and obtain the relation

Fig. 2.5 Sketch of magnetic
thin film with magnetic field
applied out of plane. The
antenna provides an
oscillating magnetic field

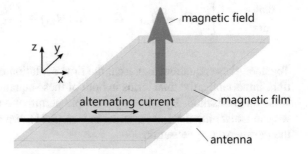

$$\frac{d}{dt} \begin{pmatrix} m_x \\ m_y \\ M_{sat} \end{pmatrix} = \gamma \begin{pmatrix} m_x \\ m_y \\ M_{sat} \end{pmatrix} \times \begin{pmatrix} -N_x m_x \\ -N_y m_y + h_y \\ -N_z M_{sat} + h_z + H_{ext} \end{pmatrix}$$

$$= \gamma \begin{pmatrix} m_y h_z + m_y H_{ext} + m_y M_{sat}(N_y - N_z) - M_{sat} h_y \\ -m_x h_z - m_x H_{ext} + m_x M_{sat}(N_z - N_x) \\ m_x m_y (N_x - N_y) + m_x h_y \end{pmatrix}. \tag{2.28}$$

Examining the RHS of Eq. (2.28), we note that there are no driving terms that are
above second order in the magnetisation. This is unsurprising, since we have assumed
that are operating at low enough amplitudes that the dynamics are linear. However,
if we increase the amplitude of $m_{x,y}$, then the corrections to the z-component of the
magnetisation must be taken into account. Regarding this, we write

$$m_z = \sqrt{M_{sat}^2 - \left(m_x^2 + m_y^2\right)}, \tag{2.29}$$

which can be expanded to second order to obtain

$$m_z = M_{sat} - \frac{1}{2M_{sat}} \left(m_x^2 + m_y^2\right). \tag{2.30}$$

With this expanded expression for m_z we note the exact origin of the nonlinearity
in this system. Since m_z now depends on the square of the transverse components,
we perform a change of variables, where every M_{sat} in Eq. (2.28) is replaced by the
RHS of Eq. (2.30). Following this transition to the nonlinear regime, and considering
only the dynamics of the transverse magnetisation (the top two rows of Eq. (2.28))
we find

$$\frac{dm_x}{dt} = \gamma \Big[m_y \big(h_z + H_{ext} + (N_y - N_z) M_{sat} \big) - M_{sat} h_y +$$

$$\frac{h_y}{2M_{sat}} (m_x^2 + m_y^2) - \left(\frac{N_y - N_z}{2M_{sat}}\right) m_y (m_x^2 + m_y^2) \Big], \tag{2.31a}$$

$$\frac{dm_y}{dt} = -\gamma \left[m_x \left(h_z + H_{\text{ext}} - (N_z - N_x) M_{\text{sat}} \right) + \left(\frac{N_z - N_x}{2 M_{\text{sat}}} \right) m_x (m_x^2 + m_y^2) \right].$$

(2.31b)

Together, these equations represent the time evolution of the transverse magnetisation. Importantly, the final terms in both of these equations are third-order in transverse magnetisation. By adopting a simple perturbation theory approach [23, 24], we can isolate the third-order terms in Eqs. (2.31). We write a power expansion of the transverse magnetisation

$$m_x = m_x^{(1)} + m_x^{(2)} + m_x^{(3)},$$

(2.32)

where $m_x^{(1)}$ is the component of m_x which is first-order in magnetisation. In addition to this, we note that for a thin-film magnetised perpendicular to the plane [25]

$$N_x \to 0 \quad N_y \to 0 \quad N_z \to 1.$$

(2.33)

Finally, by employing this approximation and the notation of (2.32) we can write

$$\frac{dm_x^{(3)}}{dt} = \frac{\gamma}{2 M_{\text{sat}}} m_y (m_x^2 + m_y^2),$$

(2.34a)

$$\frac{dm_y^{(3)}}{dt} = \frac{-\gamma}{2 M_{\text{sat}}} m_x (m_x^2 + m_y^2).$$

(2.34b)

This set of equations shows the driving terms which are the origin of third-order nonlinear effects in a perpendicularly magnetised thin-film. We shall now focus on one nonlinear effect in particular: phase conjugation via four-wave mixing.

2.3.3 Magnonic Four-Wave Mixing

The nonlinear process that we are interested in is four-wave mixing. To explore this using spin waves rather than electromagnetic waves, we consider a magnonic system shown in Fig. 2.6. The similarities between this system and that depicted in Fig. 2.4 are plain: there are two counter propagating pump waves ($m_{1,2}$), and a probe wave (m_p), that are present in a nonlinear medium. We shall now show, for the first time, that this four-wave mixing process leads to a phase conjugate reflection in a magnonic system.

We begin by defining some notation that will make the mathematics easier to handle and is consistent with the literature [23, 24]. We define the real quantity $\tilde{m}$ which is the oscillating transverse magnetisation. We express this as the sum of a complex value, m, with its complex conjugate both of which are oscillating in the Argand plane. Explicity:

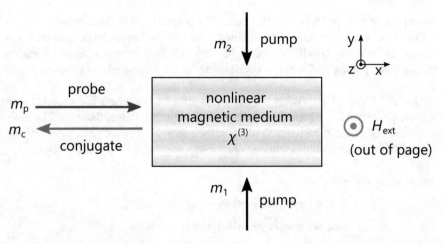

Fig. 2.6 Illustration of four-wave mixing in a magnetic system. The counter propagating pump waves mix with the incoming probe wave to create a fourth wave that is the phase conjugate of the probe wave

$$\tilde{m} = m + m^*$$
$$= ae^{i(\mathbf{k}\cdot\mathbf{r}-\omega t)} + a^* e^{-i(\mathbf{k}\cdot\mathbf{r}-\omega t)}. \tag{2.35}$$

where a is a complex amplitude. We also define the wavevectors that are involved, as illustrated in Fig. 2.6. The transverse magnetisations that are input to the nonlinear medium have wavevectors

$$\begin{aligned}
\mathbf{k}_1 &= (0, k, 0) \\
\mathbf{k}_2 &= (0, -k, 0) \\
\mathbf{k}_p &= (k, 0, 0),
\end{aligned} \tag{2.36}$$

where we have assumed the degenerate, isotropic case. In the region of the medium where all three waves overlap, the total m will be a superposition of these input waves. Explicitly,

$$\begin{aligned}
\tilde{m} =& a_1 e^{i(\mathbf{k}_1\cdot\mathbf{r}-\omega_1 t)} + a_1^* e^{-i(\mathbf{k}_1\cdot\mathbf{r}-\omega_1 t)} + \\
& a_2 e^{i(\mathbf{k}_2\cdot\mathbf{r}-\omega_2 t)} + a_2^* e^{-i(\mathbf{k}_2\cdot\mathbf{r}-\omega_2 t)} + \\
& a_p e^{i(\mathbf{k}_p\cdot\mathbf{r}-\omega_p t)} + a_p^* e^{-i(\mathbf{k}_p\cdot\mathbf{r}-\omega_p t)},
\end{aligned} \tag{2.37}$$

where the total transverse magnetisation is expressed as a real quantity as is convenient when working with nonlinear dynamics. For clarity, we have also made explicit the phase factors of each term. Furthermore, we note that with this notation, Eqs. (2.34) correspond to the statement

$$\frac{d\tilde{m}^{(3)}}{dt} \propto \tilde{m}^3, \tag{2.38}$$

where the $\tilde{m}^{(3)}$ is the transverse magnetisation proportional to third-order.

We now have a method for investigating what kind of third-order nonlinear effects can occur in the system illustrated in Fig. 2.6; that is, a system with two counter-propagating pumps and a probe overlapping in the same region of the magnetic film.

To investigate what possible driving terms arise from the multiple different spin waves, we expand $\tilde{m}^3$. Including the complex conjugate terms, there are 216 individual terms that can be reduced to 56 unique terms. After simplification and the abbreviation of the complex conjugates, the expansion is reduced to 28 terms:

$$
\begin{aligned}
\tilde{m}^3 = {} & m_1^3 + m_2^3 + m_p^3 \\
& + 3\big[m_1^2 m_1^* + m_1^2 m_2 + m_1^2 m_2^* + m_1^2 m_p + m_1^2 m_p^* \\
& \quad + m_1 m_2^2 + m_1^* m_2^2 + m_2^2 m_2^* + m_2^2 m_p + m_2^2 m_p^* \\
& \quad + m_1 m_p^2 + m_1^* m_p^2 + m_2 m_p^2 + m_2^* m_p^2 + m_p^2 m_p^* \big] \\
& + 6\big[m_1 m_1^* m_2 + m_1 m_1^* m_p + m_1 m_2 m_2^* \\
& \quad + m_2 m_2^* m_p + m_1 m_p m_p^* + m_2 m_p m_p^* \\
& \quad + m_1 m_2 m_p + m_1^* m_2 m_p + m_1 m_2^* m_p + m_1 m_2 m_p^* \big] + \text{c.c..}
\end{aligned}
$$
$$(2.39)$$

Though this expression is unwieldy, it can be pared back significantly by eliminating forbidden un-physical terms, as we shall proceed to do.

2.3.3.1 Forbidden Terms

At first sight, Eq. (2.39) appears formidable. Recall however, that each m_j contains a phase factor $e^{i(\mathbf{k}_j \cdot \mathbf{r} - \omega_j t)}$. For any one of the 28 terms to be a viable driving term, the resultant phase factor must lie within the bandwidth of the spin wave. In other words, the combination of wavenumber and angular frequency, must be supported by the magnon dispersion relation.

Let us illuminate the point with a simple example. Consider the first term in Eq. (2.39). Upon expansion, we find that

$$
\begin{aligned}
m_1^3 &= \big[a_1\, e^{i\,(\mathbf{k}_1 \cdot \mathbf{r} - \omega_1 t)} \big]^3 \\
&= a_1^3\, e^{i\,(3\mathbf{k}_1 \cdot \mathbf{r} - 3\omega_1 t)}.
\end{aligned}
$$
$$(2.40)$$

This describes a wave with a factor-of-three increase of the wavenumber and angular frequency. The dispersion relation however is not linear and so the proportional increase of k and ω corresponds to a wave that is not necessarily supported by the system. This idea is illustrated in Fig. 2.7 which shows a typical dispersion curve for the forward volume magnons that are excited in the experiment discussed in Chap. 4.

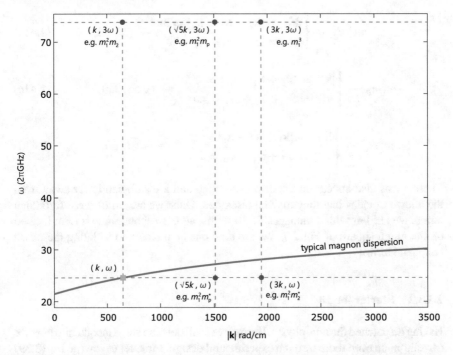

Fig. 2.7 A dispersion curve for forward volume magnetostatic spin waves, such as obtained in our experiment. The green square shows where the input magnons lie on the dispersion curve. The red circles correspond to third-order terms that are forbidden since their phase-factors put their coordinates far away from the allowed band

The green square marks the wavenumber and angular frequency for each individual input, m_j, when they are degenerate.[3] Note that the x-axis is the magnitude of $\mathbf{k}$ since $\mathbf{k}_1$ is anti-parallel to $\mathbf{k}_2$, which are both perpendicular to $\mathbf{k}_p$. This is justified since the dispersion relation is isotropic for forward volume modes and therefore has rotational symmetry, hence it is valid to plot the dispersion curve as a single line for wavevectors that point in different directions.

Regarding the phase factor of the m_1^3 term in Eq. (2.40), we can see that it corresponds to the red circle at the top right of Fig. 2.7, and is far away from the modes supported by the system, hence it may be disregarded. A similar fate applies to the second and third terms in Eq. (2.39), which also lie at the coordinate $(3k, 3\omega)$. Other examples, lying outside the bandwidth and highlighted on the graph are

$$m_1^2 m_2 \to \begin{cases} |\mathbf{k}| = |\mathbf{k}_1 + \mathbf{k}_1 + \mathbf{k}_2| = k \\ \omega = \omega_1 + \omega_1 + \omega_1 = 3\omega \end{cases} \to (k, 3\omega), \qquad (2.41a)$$

---

[3] As discussed in Chap. 4, the experiment uses degenerate pump magnons, but *non-degenerate* probe magnons. The detuning however, corresponds to approximately 1 part in 1000 and may be approximated as degenerate for this current analysis.

$$m_1^2 m_2^* \rightarrow \begin{cases} |\mathbf{k}| = |\mathbf{k}_1 + \mathbf{k}_1 - \mathbf{k}_2| = 3k \\ \omega = \omega_1 + \omega_1 - \omega_2 = \omega \end{cases} \rightarrow (3k, \omega), \qquad (2.41b)$$

$$m_1^2 m_p \rightarrow \begin{cases} |\mathbf{k}| = |\mathbf{k}_1 + \mathbf{k}_1 + \mathbf{k}_p| = \sqrt{5}k \\ \omega = \omega_1 + \omega_1 + \omega_p = 3\omega \end{cases} \rightarrow (\sqrt{5}k, 3\omega), \qquad (2.41c)$$

$$m_1^2 m_p^* \rightarrow \begin{cases} |\mathbf{k}| = |\mathbf{k}_1 + \mathbf{k}_1 - \mathbf{k}_p| = \sqrt{5}k \\ \omega = \omega_1 + \omega_1 - \omega_p = \omega \end{cases} \rightarrow (\sqrt{5}k, \omega). \qquad (2.41d)$$

These terms also appear on the dispersion map and are sufficiently far away from the allowed modes that they may be discarded. Once we are rid of these forbidden terms, we are left with a manageable 10 terms, all corresponding to a term located on the green square of Fig. 2.7. We are now one step closer to isolating the phase conjugate term.

2.3.3.2   Starter for 10

Having dispatched the non-physical terms, we shall develop some insight into how we can eliminate even more through experimental design. First, let us rewrite Eq. (2.39) in the light of the above exclusions. The remaining sum of third-order driving terms is now

$$3\left[m_1^2 m_1^* + m_2^2 m_2^* + m_p^2 m_p^*\right] + 6\left[m_1 m_1^* m_2 + m_1 m_1^* m_p + m_1 m_2 m_2^* + m_2 m_2^* m_p \right. \\ \left. + m_1 m_p m_p^* + m_2 m_p m_p^* + m_1 m_2 m_p^*\right] + \text{c.c.}. \qquad (2.42)$$

This sum however, may be reordered to emphasise the phase factor associated with each term. Rearranging the sum to collect terms of similar oscillation, we find there are 4 distinct dependences. Ordering the sum as such, we obtain

$$\begin{aligned}
3m_1^2 m_1^* + 6m_1 m_2 m_2^* + 6m_1 m_p m_p^* & \quad \Big\} \quad \propto \ e^{i(\mathbf{k}_1 \cdot \mathbf{r} - \omega t)} \\
+3m_2^2 m_2^* + 6m_1 m_1^* m_2 + 6m_2 m_p m_p^* & \quad \Big\} \quad \propto \ e^{i(\mathbf{k}_2 \cdot \mathbf{r} - \omega t)} \\
+3m_p^2 m_p^* + 6m_1 m_1^* m_p + 6m_2 m_2^* m_p & \quad \Big\} \quad \propto \ e^{i(\mathbf{k}_p \cdot \mathbf{r} - \omega t)} \qquad (2.43) \\
+6m_1 m_2 m_p^* & \quad \Big\} \quad \propto \ e^{i(-\mathbf{k}_p \cdot \mathbf{r} - \omega t)} \\
& \qquad + \text{c.c.},
\end{aligned}$$

where we have again assumed $\omega_1 = \omega_2 = \omega_p = \omega$. What Eq. (2.43) tells us is that each of the four distinct oscillations is travelling in a different direction, a notion

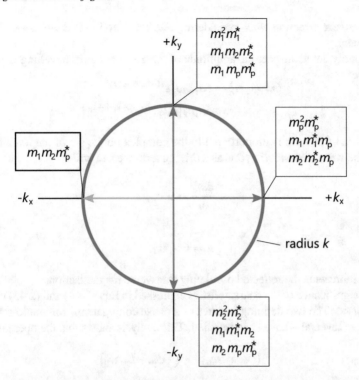

Fig. 2.8 Circle of radius $k(\omega, H_{ext})$ showing the **k**-vectors of the 10 allowed terms. Orange arrows represent the **k**-vectors of the 3 input magnons, m_1, m_2, and m_p. Boxed are terms that correspond to each of the **k**-vectors. The green arrow shows that only one term is propagating in the $-x$ direction, and that this term is proportional to m_p^*.

illustrated in Fig. 2.8. In this diagram, the orange arrows represent terms that are propagating co-linear with one of the input magnons, $m_{1,2,p}$. The presence of these terms corresponds to an amplification of the initial inputs; a common application of nonlinear dynamics.

More interestingly however, is the green arrow and the term associated with it. Firstly, notice that it is propagating in the direction of $-\mathbf{k}_p$. This may appear as a simple reflection of the wave travelling with $\mathbf{k}_p$, but we may go further and apply a stronger condition of reflection noting that as long as $\mathbf{k}_1 = -\mathbf{k}_2$, then the reflected wave will *always* have wavevector $-\mathbf{k}_p$, as is the case with a phase conjugate reflection. In other words, this reflection will always travel anti-parallel to the probe wave, or mathematically,

$$\mathbf{k}_c = -\mathbf{k}_p, \tag{2.44}$$

where $\mathbf{k}_c$ is the wavevector of the reflected wave. Furthermore, this wave is the only one that is reflected in any sense of the word, let alone the strict condition of phase conjugate reflection. This confirms that no other excited term will travel in the negative x direction when considering Fig. 2.6; a notion that will be vital to the experiment.

Secondly, let us inspect the amplitude of this reflected term travelling with $-\mathbf{k}_p$. We have:

$$6m_1m_2m_p^* = 6a_1a_2a_p^* e^{i(-\mathbf{k}_p\cdot\mathbf{k}-\omega_p t)}$$
$$= 6a_1a_2|a_p|e^{-i\phi}e^{i(-\mathbf{k}_p\cdot\mathbf{k}-\omega_p t)}. \tag{2.45}$$

We see that this term is proportional to the complex conjugate of the amplitude of the probe wave. Considering this as a driving term we can write

$$\frac{dm_c}{dt} = um_p^*, \tag{2.46}$$

where

$$u = 6m_1m_2, \tag{2.47}$$

and m_c represents the reflected oscillating transverse magnetisation.

The significance of the characteristics expressed in Eqs. (2.44) and (2.45) must be emphasised. The two defining features of a phase conjugate mirror—auto-retracing and phase reversal—have been satisfied. This analysis shows that the operation

$$(\phi_p, k_p, \omega_p) \rightarrow (-\phi_p, -k_p, \omega_p) \tag{2.48}$$

is indeed as possible for a magnonic system as it is for an optical one. Moreover, it tells us that not only is a phase conjugate term a possibility from magnon four-wave mixing, but much more significantly, it is the *only* possibility in an experiment that measures the reflection from the interaction region where the mixing occurs.

2.4   Conclusion

In this chapter, we have set out a simple theoretical framework by which a third-order nonlinear mixing process occurring in a magnetic film can, under specific conditions, result in a phase conjugate spin wave. Building from Yariv and Pepper's optical phase conjugation theory [13] and applying a similar perturbative approach to spin wave dynamics as explored by Camley [23], we obtained our key result of Eq. (2.46).

It is worth emphasising that since the theory presented above is based on a perturbation theory, it is reliant on all of the assumptions and limitations that come with such an approach as outlined in Sect. 2.3.1. This fact is paramount to any future development of this theory and creates certain challenges to be aware of when comparing with experiment. For instance, increasing the power of the probe magnons to a value comparable with the pumps will allow one to record a larger signal experimentally, but this leads to the failure of the weak probe assumption, causing other nonlinear terms to become more prominent. Another potential failure arises when driving the pumps at powers high enough to cause the onset of chaotic behaviour, leading to a linewidth broadening and the breakdown of the monochromatic approximation.

As the experimental parameters begin to stray from the regions where the assumptions are valid, a numerical approach offers a potential solution. Indeed, powerful open source software packages such as OOMMF and, as we will see in Chap. 4, MuMax3 [26] can be used to model complex magnetic systems effectively when an analytical approach is not always viable.

References

1. Shkunov VV, Zel'dovich BY (1985) Optical phase conjugation. Sci Am 253:54–59
2. Fisher RA (1983) Optical phase conjugation. Academic, New York
3. Gerritsen HJ (1967) Nonlinear effects in image formation. Appl Phys Lett 10:239–241
4. Zel'dovich BY, Popovichev VI, Ragul'skiy V, Faizullov F (1972) Connection between the wave fronts of the reflected and exciting light in stimulated Mandel'shtem-Brillouin scattering. JETP Lett 15:109
5. Yariv A (1976) On transmission and recovery of three-dimensional image information in optical waveguides. J Opt Soc Am 66:301
6. Kawakami K, Komurasaki K, Okamura H (2017) Retrodirective tracking of a moving target using phase conjugate light generated in a Fabry-Pérot Nd:YAG laser. J Appl Phys 121:093104
7. Chiou AE (1999) Photorefractive phase-conjugate optics for image processing, trapping, and manipulation of microscopic objects. Proc IEEE 87:2074–2085
8. Unnikrishnan G, Joseph J, Singh K (2008) Optical encryption system that uses phase conjugation in a photorefractive crystal. Appl Opt
9. Boyd RW (2008) Nonlinear optics, 3rd edn
10. Jackson JD (1998) Classical electrodynamics, 3rd edn
11. Hellwarth RW (1977) Errata: generation of time-reversed wave fronts by nonlinear refraction. J Opt Soc Am 68:1155
12. Bloom DM, Bjorklund GC (1977) Conjugate wave-front generation and image reconstruction by four-wave mixing. Appl Phys Lett 31:592–594
13. Yariv A, Pepper DM (1977) Amplified reflection, phase conjugation, and oscillation in degenerate four-wave mixing. Opt Lett
14. Gordon AL et al (1998) Phase conjugation of linear signals and solitons of magnetostatic waves. JETP Lett 67:913–918
15. Melkov GA, Sholom SV (1991) Kinetic instability of spin waves in thin ferrite films. Sov Phys JETP 72:341–346
16. Melkov G, Serga A, Tiberkevich V, Oliynyk A, Slavin A (2000) Wave front reversal of a dipolar spin wave pulse in a nonstationary three-wave parametric interaction. Phys Rev Lett 84:3438–41
17. Melkov G, Kobljandkyj Y, Serga A, Tiberkevich V, Slavin A (2001) Nonlinear microwave signal processing in yttrium-iron garnet (YIG) films. In: European Conference on Circuit Theory and Design, pp I-293–296
18. Serga AA et al (2005) Parametric generation of forward and phase-conjugated spin-wave bullets in magnetic films. Phys Rev Lett 94:167202
19. Melkov GA, Vasyuchka VI, Chumak AV, Slavin AN (2005) Double-wavefront reversal of dipole-exchange spin waves in yttrium-iron garnet films. J Appl Phys 98
20. Melkov GA, Koblyanskiy YV, Slipets RA, Talalaevskij AV, Slavin AN (2009) Nonlinear interactions of spin waves with parametric pumping in permalloy metal films. Phys Rev B 79:1–9
21. Vasyuchka VI et al (2010) Non-resonant wave front reversal of spin waves used for microwave signal processing. J Phys D Appl Phys 43:325001

22. Butikov EI (2004) Parametric excitation of a linear oscillator. Eur J Phys 25:535–554
23. Marsh J, Camley RE (2012) Two-wave mixing in nonlinear magnetization dynamics: a perturbation expansion of the Landau-Lifshitz-Gilbert equation. Phys Rev B 86:224405
24. Khivintsev Y et al (2011) Nonlinear amplification and mixing of spin waves in a microstrip geometry with metallic ferromagnets. Appl Phys Lett 98
25. Osborn JA (1945) Demagnetizing factors of the general ellipsoid. Phys Rev 67:351–357
26. Vansteenkiste A et al (2014) The design and verification of MuMax3. AIP Adv 4:107133

Chapter 3
Experimental Methods and Details

Yttrium-iron garnet... is a marvel of nature. Its role in the physics of magnets is analogous to that of germanium in semiconductor physics, water in hydrodynamics, and quartz in crystal acoustics

Victor L'vov The Saga of YIG [1]

In this chapter we present a detailed background of the experiment serving as the core of this thesis. We introduce the materials used, followed by the design and characterisation of vital antennae. Latterly, equipment central to the experiments of Chaps. 4 and 5 is described.

3.1 Yttrium Iron Garnet

One of the most commonly used materials in magnonics is yttrium iron garnet (YIG), $Y_3Fe_5O_{12}$. YIG has been of scientific interest for well over 60 years since rare-earth iron garnets—the group to which YIG belongs—were first discovered [2, 3]. It is a material used extensively in microwave applications [4], chiefly because of its very narrow resonance linewidth, allowing one to achieve resonators with exceptionally high Q values.

YIG is a ferrimagnetic insulator composed of two antiferromagnetically coupled sublattices. Each unit cell contains 4 unit formulae totalling 80 atoms. The crystal has a cubic structure with a lattice constant $a_0 = 12.376\,\text{Å}$. The magnetism arises from the 20 Fe^{3+} ions ($J = S = 5/2$, $g = 2$, $m = 5\,\mu_B$) which are divided into two antiferromagnetically coupled sublattices with 12 Fe^{3+} ions in a tetrahedral O^{2-} environment and 8 in an octahedral O^{2-} environment. A section of the unit formula dipicting the two Fe^{3+} configurations is depicted in Fig. 3.1.

A. Inglis, *Investigating a Phase Conjugate Mirror for Magnon-Based Computing*, Springer Theses, https://doi.org/10.1007/978-3-030-49745-3_3

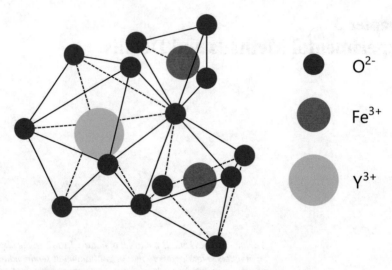

Fig. 3.1 Yttrium iron garnet ($Y_3Fe_5O_{12}$, YIG) is an insulating ferrite. It has two antiferromagnetically coupled sublattices of Fe^{3+} ions. The large unit cell of 12.376 Å houses 80 atoms, 20 of which are Fe^{3+} ions. Two different O^{2-} environments in the unit cell define the Fe^{3+} sublattices: 8 Fe^{3+} ions are in a tetrahedral environment and 12 are in an octahedral environment. A section of the unit formula with examples of both O^{2-} environments is shown

The enduring popularity of YIG is largely a consequence of the exceptionally low damping parameter in comparison to other magnetic materials. This low damping is in part due to a relatively slow spin-lattice relaxation arising from the magnetic Fe^{3+} ions being in the spherically symmetric state $^6S_{5/2}$ [5]. An important factor contributing to the reduced damping is the insulating nature of YIG. The absence of delocalized electrons eliminates two damping mechanisms that are present in metallic magnetic waveguides: internal eddy current damping[1] [6, 7], and radiation shielding [8, 9]. In addition to its pervasive use in high quality microwave components, the long magnon lifetime offered by YIG has positive consequences for experiments involving dipolar magnons. Typically, the wavelengths of dipolar spin waves in YIG thin films are in the micron to millimetre range, while their propagation over centimetre scales is readily observable.

This long relaxation length is the main reason that YIG was used for the magnon waveguide as it allows for experiments to be performed on the millimetre scale commensurate with our waveguide and excitation methods (see Sect. 3.3). The waveguide employed in the experiments described in this thesis is approximately 18 mm

---

[1]It is worth noting that when considering a conducting magnetic waveguide there are two channels by which eddy currents can dissipate energy from the precessing spins. There are the eddy currents in the waveguide itself (as alluded to above), but there are also eddy currents in the metallic antenna or coplanar waveguide used to excite the spin waves. The latter type of damping is referred to historically as radiative damping and is dependent on the excitation method. In general, there are many damping mechanisms and isolating the contribution intrinsic to the material is experimentally challenging.

long, 2 mm wide, and 7.8 μm thick. It is a high-quality, monocrystalline film grown using the method of high-temperature liquid-phase epitaxy on a substrate of gallium gadolinium garnet ($Gd_3Ga_5O_{12}$,GGG). GGG is commonly used as a substrate since it has a lattice constant of 12.383 Å, similar to that of YIG, thereby minimising stress from lattice mismatch.

With its fairly high Curie temperature ($T_C = 559$ K), YIG is placed in good stead for use in any future room temperature wave-computing devices. As the temperature tends towards 0 K, the saturation magnetisation of YIG tends towards a maximum of approximately $M_{sat} = 196$ kAm^{-1}. Since all experiments were performed at room temperature however, we use the reduced value of $M_{sat} = 140$ kAm^{-1} in any calculations present throughout.[2]

3.2 Exciting and Detecting Magnons

There a number of ways to excite and detect spin waves experimentally. The differing methods each offer their own benefits and drawbacks and an awareness of these is important for any magnonic practitioner.

3.2.1 Spin Pumping and Spin-Transfer Torque

In recent years there has been a great interest in the interconversion between the spin angular momentum of a conduction electron and that of a precessing magnetisation. This transference of spin can be performed in either direction, though each process has a different name historically. The term 'spin pumping' refers to the conversion of the spin angular momentum carried by a spin wave into the spin state of a conduction electron [10, 11]. An interesting aspect of spin pumping is that it provides a way of transmitting spin information from a medium that is necessarily magnetically ordered to one that need not have any long range magnetic order. A particularly pleasing example of this is spin pumping from YIG, an insulating ferrimagnet, into Pt, a conducting paramagnet [12]. Spin-transfer torque (STT), is the complementary process to spin pumping. The magnetisation of the material absorbs spin from a conduction electron that is reflected at the boundary whilst simultaneously undergoing a spin-flip [13–15]. In order for this to result in the excitation of a spin wave however, there must be a net spin current in the material providing the conduction electrons (Fig. 3.2).

---

[2]As a truly remarkable material, there have been many studies of YIG over the years. A interested reader requiring a concise table of physical and magnetic properties of YIG should be directed towards Appendix A of *Spin Waves* by Stancil and Prabakhar [5], while those looking for a more technically and theoretically involved description of magnon behaviour in YIG should see the thorough review by Cherepanov *et al.* [1].

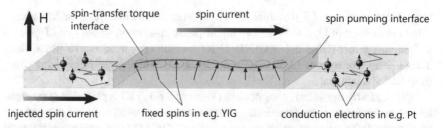

Fig. 3.2 A schematic illustration of two types of spin current in solids and the interconversion between them. In the edge materials (orange) the spin is carried via electron diffusion, while in the middle material (pink) the spin current is transmitted via the collective precession of magnetized spins. The left interface shows where spin-transfer torque facilitates the conversion of diffusive spin current to magnon spin current, while the right-most interface shows the creation of a diffusive spin current from spin pumping

These processes represent the interchange between spin waves and spin current. Over the past few years, the biggest challenge to experiments is the handling of spin currents; creating pure spin currents large enough to excite spin waves via STT, and detecting small spin currents generated by spin pumping. Many experiments exploit an attribute intrinsic to heavy metals, such as Pt or W, which is a coupling between spin current and charge current called the spin Hall effect (SHE) and the inverse spin Hall effect (ISHE). The strength of coupling between spin and charge currents is quantified by a parameter known as the spin Hall angle, θ_{SH}, which varies for different materials[3] Recently however, experiments have managed to harness the large pure spin currents that are found in topological insulators [23].

The main advantage of this technique is that it allows for the excitation of a spin wave using a DC current, while being able to detect magnons using a sensitive voltmeter [22]. The use of DC current however, will result in Joule heating which is problematic for a low-power wave-computing paradigm (Fig. 3.3).

3.2.2 Phonon-Magnonics

The foundations of coupling between phonons and magnons has been studied for many decades, starting with the study of magnetoelastic behaviour. There has been increased interest in the past decade with the development of potential devices to excite spin waves via acoustic waves [24–29]. One of the main advantages of acoustic excitation of spin waves is voltage driven piezoelectric crystals that operate at comparatively low-power and eliminate Joule heating effects. The theoretical model

---

[3]Examples of experiments involving the detection of magnons using ISHE can be found in Refs. [16–18]. Readers interested in aspects of the SHE such as its magnitude and various dependancies are guided towards Refs. [16, 19, 20]. A useful comparison of spin Hall angles for various metals can be found in Ref. [21] while an impressive collaborative effort to determine the absolute sign of the spin Hall angle is found in Ref. [22].

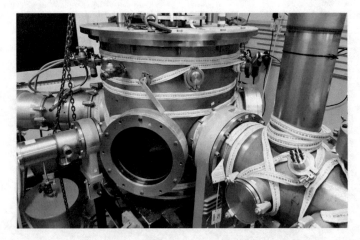

Fig. 3.3 A vacuum chamber used for deposition of Pt films onto YIG which were then used to detect inverse spin Hall voltages

describing the coupling between magnons and phonons, relies on the magnetic free energy density [30] and involves a term that is proportional to the product of the strain induced by an elastic wave and the specific magnetoelastic coupling for the given material.

Recently, a novel device for exciting BVMSWs using a longitudinal standing acoustic wave was developed [31]. The device, shown in Fig. 3.4, utilizes a YIG waveguide on a GGG substrate. On the GGG-air interface, a ZnO transducer is grown which launches an elastic wave of GHz frequency into the YIG film which causes an oscillating strain that couples to a backward-volume spin wave mode of the same frequency within the YIG. The images shown in (b) are of devices grown in-house and show multiple transducer discs fabricated on the waveguide.

3.2.3 EM Coupling

One of the first reports of collective spin precession was in 1946 when Griffiths excited ferromagnetic resonance using an electromagnetic cavity resonator [32]. Of the excitation methods described, this coupling is historically the most widespread and is arguably the most intuitive to the passing physicist. In this method, spin waves are excited inductively, whereby the ordered spins—aligned with the large static external magnetic field—are exposed to a much smaller perpendicular oscillating magnetic field. In trying to align themselves with this oscillating field, the spins are excited away from the fully ordered ground state (Fig. 3.5).

Exciting magnons in this way is remarkably straightforward and can be done with a simple antenna. A thin conducting wire placed in the vicinity of the waveguide will

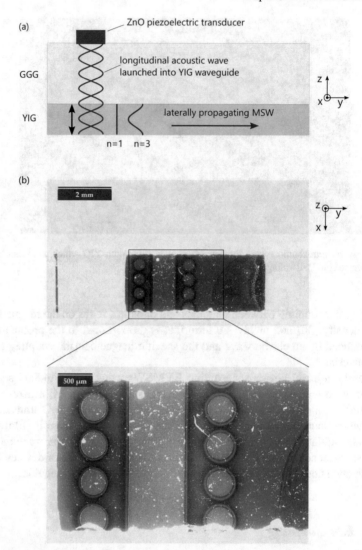

Fig. 3.4 Sketch (**a**) and images (**b**) of novel phonon magnonic device for exciting spin waves. The sketch shows a side-on view ZnO piezoelectric transducer placed on the back-side of the GGG substrate. Longitudinal acoustic waves propagate into the YIG waveguide and are reflected from the boundary. The standing acoustic wave couples to different magnon thickness modes (shown in blue) via the magnetoelastic effect. The images shown in (**b**) are a top-down view of devices grown in-house, where each of the circles is a ZnO transducer

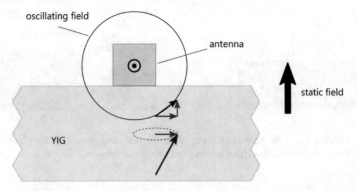

Fig. 3.5 Side-view schematic of excitation antenna on YIG waveguide. The excitation is shown for the forward volume process whereby in the absence of antenna current, the magnetisation will be saturated upwards (out of the plane of the film). As a microwave current passes through the antenna, the horizontal component of the induced field exerts a time-dependent torque on the local spins, exciting localized precession which can—under the correct conditions—give rise to a propagating spin wave

excite a spin wave if a current—oscillating at a frequency supported by the passband of the waveguide—is passed through the wire. It is precisely because of the simplicity with which magnons can be inductively excited and detected that the experiments described in Chaps. 4 and 5 utilize such antennae.

3.3 Antenna Design

In this section we expand upon the introduction to antenna excitation above, and discuss the process of designing the antennae used in the experiments detailed in Chaps. 4 and 5. Above, the workings of the excitation of spin waves using an antenna were only touched upon, with an appeal to intuition. There is a vast amount of literature[4] demonstrating the validity of this method for the excitation and detection of magnons. Detailed analyses of the coupling between antenna fields and magnon modes are presented in Refs. [40, 41], while a calculation of the antenna field distribution is given in Ref. [42]. Only the main results of these analyses are presented here. We note that the amplitude $m_{n,i}(k_y)$, of the nth thickness mode of a spin wave travelling in the y-direction with wavenumber k_y in an out-of-plane (z-direction) magnetized waveguide excited by the i-component of the magnetic field induced by the antenna is

$$m_{n,i}(k_y) = |\tilde{b}_{z,n}||\tilde{b}_{i,y}(k_y)|\left[\frac{f_n(k_y)}{\gamma} + \frac{1}{M_{\text{sat}}}\left(\mu_0 H_{\text{ext}}^2 - \frac{f_n(k_y)^2}{\gamma^2}\right)\right], \quad (3.1)$$

---

[4]For a few experiments where this standard technique is exploited, see for example [32–39].

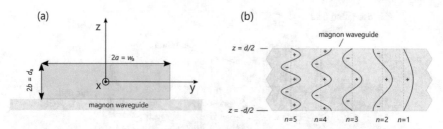

Fig. 3.6 **a** Illustration of the antenna geometry where $2a = w_a$ and $2b = d_a$. **b** Sketch of amplitude profile of thickness modes throughout the magnon waveguide for $n = 1, 2, 3, 4, 5$

where damping and ellipticity of precession have been ignored. Here, $|\tilde{b}_{z,n}|$ represents the convolution of the magnon nth thickness mode profile with the antenna field along the z-direction, $|\tilde{b}_{i,y}(k_y)|$ represents the 1D Fourier transform of the i-component of the antenna field $\tilde{b}_i(y)$ with the y-direction as the spatial domain. The frequency of the nth thickness mode magnon propagating with wavenumber k_y is given by $f_n(k_y)$, while the remaining γ, H_{ext}, M_{sat} are the familiar gyromagnetic ratio, externally applied magnetic field, and saturation magnetisation, respectively.

We can see from Eq. (3.1) that the amplitude of the excited spin wave not only depends on the dispersion characteristics (given in the square brackets), but also on the geometry of the antenna field and how this overlaps with the spin wave magnetisation profile. A typical operating frequency for the antenna field is 3 GHz, corresponding to a free space wavelength of approximately 10 cm, allowing us to assume that the field is uniform over the thickness of the film.

If we assume that the thickness modes have a cosine-shaped profile across the thickness of the film from $z = -d/2$ to $d/2$, as in Fig. 3.6b, then the nth thickness mode will have the form

$$m_n(z) \propto \cos\left(\frac{n\pi z}{d}\right). \tag{3.2}$$

Since we have assumed the antenna field is constant throughout the thickness of the waveguide, the convolution with the mode profile collapses to a simple integral of the mode profile over the waveguide thickness, multiplied by the constant field. Therefore, we have

$$|\tilde{b}_{z,n}| \propto \left| \int_{-d/2}^{d/2} \cos\left(\frac{n\pi z}{d}\right) dz \right| = \left| \frac{2d}{n\pi} \sin\left(\frac{\pi n}{2}\right) \right| = \begin{cases} 0, \ n \ \text{even} \\ \frac{1}{n}\frac{2d}{\pi}, \ n \ \text{odd} \end{cases}. \tag{3.3}$$

We see from this expression that for a uniform antenna field, even modes cannot be excited—a result established by Rosenberg and Phillips [43]. Furthermore,[5] modes that are odd in n scale as $1/n$ and higher order modes quickly become negligible.

---

[5]Note that we have calculated that the *amplitude* scales as $1/n$, while the quantity that is usually measured, the *intensity*, scales like $1/n^2$.

Examining Eq. (3.1) further, we consider the term $|\tilde{b}_{i,y}(k_y)|$, as this also depends on the antenna field geometry. Since this term represents the Fourier transform of the field, the spatial field distribution must first be calculated. Using the antenna geometry in Fig. 3.6a, and the main result from Ref. [42], we find that the y and z components of the field are given by:

$$
\mu_0 h_y(y,z) = -\frac{I\mu_0}{8\pi ab}\Bigg\{(a-y)\Bigg[\frac{1}{2}\ln\left(\frac{(b-z)^2+(a-y)^2}{(-b-z)^2+(a-y)^2}\right)+\frac{b-z}{a-y}\arctan(\frac{a-y}{b-z})
$$
$$
-\frac{-b-z}{a-y}\arctan(\frac{a-y}{-b-z})\Bigg]-(-a-y)\Bigg[\frac{1}{2}\ln\left(\frac{(b-z)^2+(-a-y)^2}{(-b-z)^2+(-a-y)^2}\right)
$$
$$
+\frac{b-z}{-a-y}\arctan(\frac{-a-y}{b-z})-\frac{-b-z}{-a-y}\arctan(\frac{-a-y}{-b-z})\Bigg]\Bigg\}
$$

(3.4a)

and,

$$
\mu_0 h_z(y,z) = -\frac{I\mu_0}{8\pi ab}\Bigg\{(b-z)\Bigg[\frac{1}{2}\ln\left(\frac{(b-z)^2+(a-y)^2}{(b-z)^2+(-a-y)^2}\right)+\frac{a-y}{b-z}\arctan(\frac{b-z}{a-y})
$$
$$
-\frac{-a-y}{b-z}\arctan(\frac{b-z}{-a-y})\Bigg]-(-b-z)\Bigg[\frac{1}{2}\ln\left(\frac{(-b-z)^2+(a-y)^2}{(-b-z)^2+(-a-y)^2}\right)
$$
$$
+\frac{a-y}{-b-z}\arctan(\frac{-b-z}{a-y})-\frac{-a-y}{-b-z}\arctan(\frac{-b-z}{-a-y})\Bigg]\Bigg\},
$$

(3.4b)

where I is the current flowing through the antenna, while a and b represent the half-width and half-height of the antenna, respectively. It is worth noting that the field distributions in Eqs. (3.4) were calculated from a simple Biot-Savart model with the assumption of a DC current. Numerical calculations have been performed however, that show good agreement between the DC-generated fields and the time-averaged AC-field distribution, making this a fair approximation [44]. An example of the magnetic field distribution calculated using Eqs. (3.4) can be found in Fig. 3.7a. The single antenna, depicted in grey, is surrounded by a vector field representing the magnetic field strength and direction calculated for an into-page current passing through an antenna with a height of 20 μm and width of 40 μm. Given this field distribution, we are now in a position to calculate the corresponding $|\tilde{b}_{i,y}(k_y)|$ term, illustrating which wavevectors are most likely to be excited for a given geometry. Figure 3.7b is the Fourier transform of the y-component of the antenna field in the y-direction. The field that was transformed is located along the dashed line in Fig. 3.7a, 2 μm below the antenna. The peak centred around zero shows that during a sweep of the external biasing field, the most efficiently excited spin waves will be the $k = 0$ mode, or ferromagnetic resonance (FMR).

As alluded to in Chap. 2 and discussed in detail in Chap. 4, the four-wave mixing experiment requires the creation of a 'pumped' region, containing a much larger amplitude of magnons than the rest of the waveguide. Because the experiment relies on spatially distinct waveguide regions, it was important to design antennae that

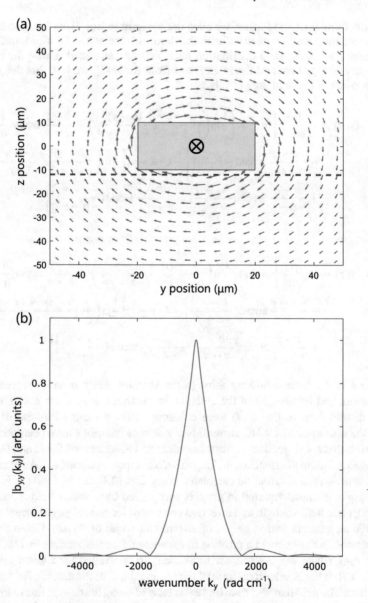

Fig. 3.7 a Magnetic field distribution calculated using Eqs. (3.4) for single antenna (grey) with a width of 40 μm and height of 20 μm. Current I is flowing in the $-x$ direction, into the page. **b** Relative amplitudes of the different spatial frequency components of the antenna field, obtained by performing a Fourier transform of the y-component of the induced field along the y-direction at a distance of 2 μm from the bottom of the antenna as depicted by the dashed line in (**a**)

would suppress the excitation of long wavelength modes for multiple reasons outlined below.

1. **Diffraction.** The pump antennae lie along part of the length of the waveguide, on opposite edges to one another. To improve confinement of the pump magnons to the region subtended by the pump antennae and so to reduce the likelihood of leakage down the rest of the waveguide, it was desirable to suppress long wavelength magnons that are more prone to diffraction.
2. **Not propagating.** The collective precession of the FMR mode is, by definition, the $k = 0$ mode. It is therefore not propagating as the excitation is coherent everywhere in the waveguide. Exciting spin waves with large wavelengths on the order of the waveguide—and thus with small momentum—means by Heisenberg's uncertainty principle, that it becomes impossible to attribute any effect to one region of the waveguide since the waves exist everywhere.
3. **Nonlinearities.** YIG is a highly nonlinear system. When the system is flooded with $k \approx 0$ magnons, this opens the door to a number of nonlinear processes. For example, the third order nonlinear process where two FMR magnons can scatter into counter propagating magnons with finite wavenumber. This process conserves energy and momentum. An illustrative sketch of this is shown in Fig. 3.8. These decays are allowed, because of the finite bandwidth of the waveguide.

The coupling to small k magnons was reduced by breaking the symmetry of the antenna. It is clear in Fig. 3.7a that the y-component of the field in the y-direction does not change sign along the dashed line, or indeed anywhere that would lie within the waveguide of thickness 7.8 µm. This of course means that there is a large component that is constant, corresponding to a peak centred around $k = 0$ in the Fourier transform. This symmetry was broken by changing the antenna from a simple wire to a meander pattern, illustrated in Fig. 3.9. By introducing a spatially periodic pattern, the large $k = 0$ coupling is suppressed, and the strongest coupling is to spin waves with finite k.

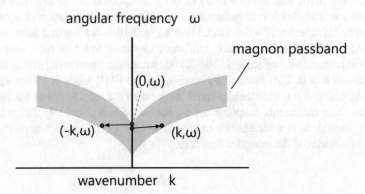

Fig. 3.8 Sketch showing an example of a magnon scattering process. Two FMR magnons with zero momentum and combined energy $2\hbar\omega$ decay into two magnons with opposite and finite momentum. The energy of each magnon remains the same as they scatter into the allowed dispersion band

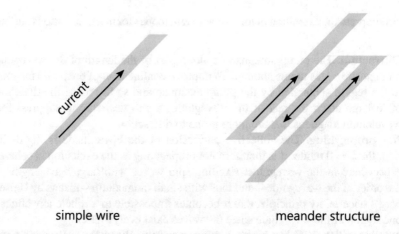

simple wire   meander structure

Fig. 3.9 High-symmetry simple wire antenna, compared with the lower-symmetry meander structure. Importantly, the meander antenna introduces a degree of periodicity, reducing the coupling to the FMR mode

To investigate the effects of breaking the symmetry of the simple wire antenna, the model described by Eqs. (3.4) was adapted for the meander structure. This was done by treating the meander as three distinct wires, where current flows in opposite directions in neighbouring wires. This approximation is valid if the length of each 'leg' of the meander is large compared to the waveguide dimensions. This is the case for the probe antenna, but is not for the pump antennae, the effects of which are discussed in a later chapter. An example of the field distribution for this arrangement can be found in Fig. 3.10a. The figure shows the field configuration for three antennae legs, each with a width of 40 μm and height of 20 μm spaced 40 μm apart, with equal current flowing through them. The central leg has current flowing into the page, while the outer two have current flowing out of the page, as illustrated in Fig. 3.9. We can see from Fig. 3.10a that the symmetry of the y-component of the field surrounding the central antenna leg is now reduced. Considering the field variation along a depth relevant to the experiment—the dashed line in Fig. 3.10a—we see that there are some higher spatial frequencies, and any uniform component has now been suppressed. This is demonstrated explicitly in Fig. 3.10b, where the Fourier transform shows a minimum at $k = 0$. This means that coupling to the FMR mode is eliminated, and is greatly reduced for wavelengths much larger than the total width of the meander structure. The maximum coupling occurs for $|k|$ of approximately $340\,\mathrm{rad\,cm^{-1}}$, which corresponds to a wavelength of approximately 185 μm, which is on the order of the total width of the meander structure.

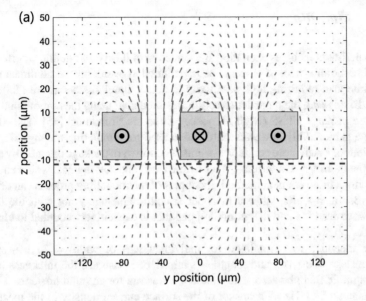

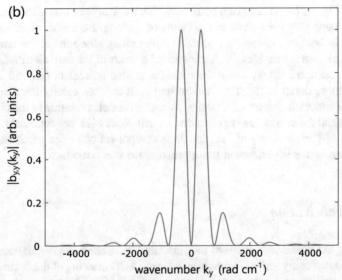

Fig. 3.10 a Magnetic field distribution calculated for antenna (grey) with meander structure. Each leg has a width of 40 μm and height of 20 μm. The central antenna leg has current flowing into the page, while the outer legs have current flowing out of the page. The dashed line marks the plane for which the Fourier transform of the field is calculated. The relative amplitude of the spatial frequencies of the y-component of the field in the y-direction for the meander structure (**b**) show a minimum at the $k = 0$ FMR mode

3.3.1 Modelling

The experiment involves excitation and detection of third order nonlinear effects and therefore requires a lot of power to be transmitted to overcome the nonlinear threshold while also requiring high sensitivity to detect small signals above the thermal noise. To this end, the meander structure was modelled using the commercial Ansys software package HFSS (High Frequency Structure Simulator) [45]. The software package is incredibly powerful and is often used for simulating the high frequency behaviour of much more complex systems. The software package operates by accepting some boundary conditions (input by the user) and solving Maxwell's equations [46] using a finite element method. Examples of the modelling process can be found in Fig. 3.11a, b which show a meander structure antenna along with the tapered microwave feedline and the grounding pad that is capacitively coupled to electrical ground.

The meander was simulated for individual legs of width $= 20 \mu$m, spacing between legs $= 30 \mu$m, and length $= 4$ mm. HFSS assumes the thickness of conductors on PCBs to be zero. The simulation was run for an input power of 1 W and a frequency of 4 GHz. A snapshot of the surface current density in the meander is shown in Fig. 3.11a. At the given phase at which the snapshot was taken, the current density changes by approximately a factor of 3 along the length of the meander. While, it is desirable to have a constant current along along the entire length, this level of variation is manageable. A transparent overlay of the field distribution taken at a cross-section halfway down the meander is also present in (a), and is shown in much more detail in (b). The field around each antenna clearly has a rotational quality, as expected. We note that the outer legs have fields rotating in the clockwise direction, and the central leg is generating an anti-clockwise magnetic field. It is this reduction of the symmetry of the transverse component of the magnetic field close to the antenna that is paramount to suppressing the $k = 0$ mode.

3.3.2 Fabrication

Following the promising results of the simulated antennae structure, the design was then fabricated using a chemical etching process. A 2D drawing of the desired design was generated using the software package AutoCAD [47]. This was printed in high-resolution on an acetate film to create a photomask. The antennae were made on printed circuit boards (PCBs) by the photofabrication unit at the University of Oxford, Department of Physics. The material onto which the circuits were printed was a board of FR-4[6] with a thin layer of copper laminated onto both sides. The circuit designs

---

[6]This is an incredibly popular composite material when it comes to PCBs. There are many reasons for this popularity, such as having a high mechanical strength, being very lightweight, and possessing extremely low water absorptivity, to name a few. It is manufactured from a fibreglass weave that is set with an epoxy resin.

(a)

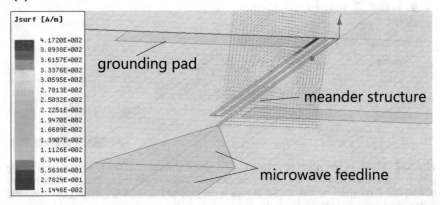

(b)

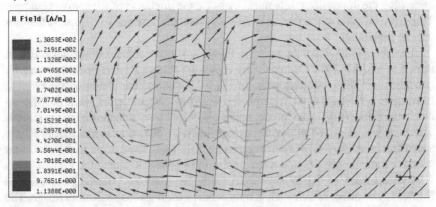

Fig. 3.11 HFSS simulations of meander antenna with individual leg width = 20 μm and spacing = 30 μm. Thickness of surface conductors is assumed zero in HFSS. The active region of the antenna is powered by the microwave feedline and is grounded capacitively with a grounding pad at the end of the meander. A map of the surface current density at a specific time is isolated for the meander only. The cross-section of the field distribution which is overlain in (**a**) is magnified in (**b**) where the broken symmetry of the field distribution is clear

were printed onto only one side of the board, with the other side acting as a ground plane. The FR-4 board was 1.5 mm thick, while the copper laminate was 18 μm thick. When the thickness of the laminate conductor is of the same order as the dimensions of the features to be printed, an effect referred to as 'side-etch' must be taken into account. This process, illustrated in Fig. 3.12, is when the sides of a desired feature are eroded by the etchant: as more of the naked copper is removed, the sides of the feature become exposed and are free to react. A general working rule for designing a photomask, is that one should compensate for this side-etch process by

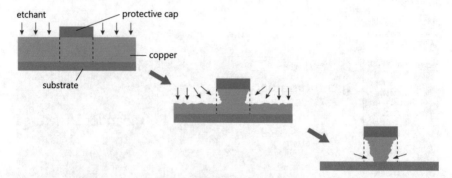

Fig. 3.12 Sketch of the effect of side-etch. The blue cap represents the protective layer in the pattern of the desired feature. The bare copper (orange) is attacked by the etchant. As the top layers of copper are removed the side of the feature is exposed to the etchant, resulting in a narrower feature than designed in the photomask

increasing the dimensions of features by twice the thickness of the copper cladding. This process is one of the factors that limits precision to which microscopic features may be fabricated.

A working PCB after fabrication is shown to varying degrees of magnification in Figs. 3.13 and 3.14. The former clearly shows the microwave feedlines, the meander structures and the grounding pads. In Fig. 3.13, the two parallel antennae running from top to bottom act as the 'pumps' and were placed 2 mm apart, while the perpendicular structure (top) is the 'probe' and is offset by 1 mm from the end of the pumps. This PCB was designed with meander legs that were 50 μm wide and spaced 30 μm apart. The boxed region in (a) is magnified and shown in more detail in (b). The latter contains a small boxed region which is shown in much more detail in Fig. 3.14.

A comparison of the finished PCB, with the AutoCAD drawing is shown in Fig. 3.14. The effect of side-etch is noticeable in the high-magnification image. The design has a width of 50 μm and a spacing of 30 μm, while the image clearly differs from this in a way that is not uniform. For example, the central leg is 40 μm wide, while the outer legs are only 28 μm wide. Indeed, the feedline at its narrowest point is only 16 μm, which is less than the thickness of the copper cladding. The different rates of side-etch can be explained by how accessible the given feature is: the central leg is partially shielded on both sides by the outer legs, which are in turn only shielded on the side closest to the centre, while the feedline is fully exposed on both sides and suffers the most erosion. The narrow feedline is a good example of the rule of thumb when designing photomasks; that the feature after fabrication will be smaller than the design by twice the thickness of the copper laminate: $50 \, \mu m - 18 \, \mu m - 18 \, \mu m = 14 \, \mu m \approx 16 \, \mu m$.

(a)

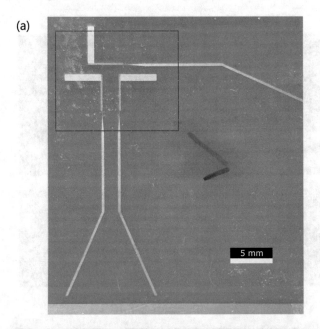

(b)

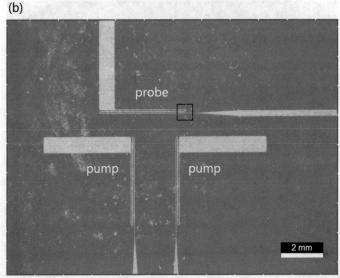

Fig. 3.13 Images of PCB after fabrication. The dashed box in (**a**) is shown under magnification in (**b**). The feedlines, meanders and grounding pads are all visible. The two parallel antennae act as pumps, while the single antenna on top is the probe. The region inside the small box in (**b**) is shown in much more detail in Fig. 3.14

(a)

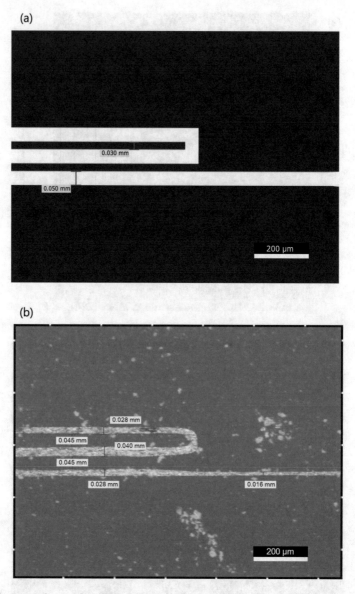

(b)

Fig. 3.14 Comparison of AutoCAD design with fabricated PCB. The design has tracks of width 50 μm and spacing of 30 μm, while the PCB has varying dimensions due to the effects of side-etch

3.4 Antenna Behaviour

Following the fabrication of the antennae, tests of their efficiency were performed. The field distribution was calculated for the antenna dimensions measured in Fig. 3.14b. The calculated field distribution was then Fourier transformed to determine the best wavevector coupling for the antenna, the results of which are shown in Fig. 3.15. Also shown is the dispersion of the first 5 FVMSW thickness modes in a thin film of thickness $d = 7.8\,\mu$m, and $M_{sat} = 140\,$kAm^{-1}, and applied field $H = 3050\,$Oe. The graph shows clearly that for significant excitation of the $n = 1$

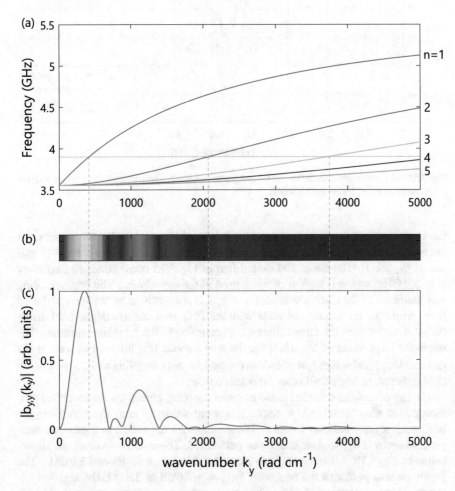

Fig. 3.15 **a** FVMSW thickness modes for $n = \{1, 2, 3, 4, 5\}$ in a film with $d = 7.8\,\mu$m, $M_{sat} = 140\,$kAm^{-1}, and $H = 3050\,$Oe. The horizontal dotted line at 3.9 GHz shows a typical experimental excitation frequency. **b** Colourmap of (**c**) which shows $|\tilde{b}_{y,y}(k_y)|$, the Fourier transform of the field distribution for the fabricated antenna, depicted in Fig. 3.14

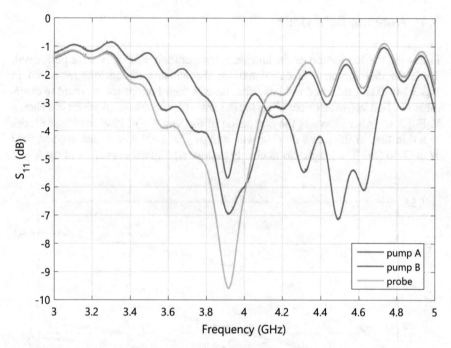

Fig. 3.16 S_{11} measurements for the probe and pump antennae. The transmission profiles all have a minimum of reflection around 3.9 GHz

thickness mode to occur (i.e. significant values of $|\tilde{b}_{y,y}(k_y)|$), the wavenumber of the next thickness mode at the same frequency corresponds to a vastly reduced values of $|\tilde{b}_{y,y}(k_y)|$. The horizontal dotted line in Fig. 3.15 corresponds to frequency $f = 3.9$ GHz, which is typical of those used in the experiment. The corresponding wavenumbers of the thickness modes $n = 1, 2, 3$ are signified by the vertical dotted lines, while the intersection of these with the Fourier transform of the field distribution demonstrates the large difference in coupling to the fabricated antenna. The relatively large value of $|\tilde{b}_{y,y}(k_y)|$ for the $n = 1$ mode (for $|k| > 0$) as well as the fact that $|\tilde{b}_{z,n}|$ scales like $1/n$ allows us to consider only coupling to the $n = 1$ mode as significant, to neglect all other thickness modes.

Having investigated the inductive magnon coupling efficiency, the electrical efficiency was also tested. SMA connectors were soldered onto the boards at the beginning of each antenna feedline. These were then connected to a network analyser, whereby a S_{11} measurement was performed. The results obtained are shown below in Fig. 3.16. All antennae have a minimum value of S_{11} around 3.9 GHz. The probe antenna performs the best with $S_{11\mathrm{pr}} = -9.6$ dB at 3.918 GHz, corresponding to a power reflection of 11%. The worst performing antenna was pump B with $S_{11\mathrm{B}} = -5.7$ dB (27% reflection) at 3.915 GHz, while pump A had $S_{11\mathrm{A}} = -7.0$ dB (20% reflection) at 3.912 GHz.

3.5 Equipment

Both experiments described in Chaps. 4 and 5 share certain experimental instruments. The following section will serve as a guide to the key items.

3.5.1 Electromagnet

As the behaviour of magnons is central to this thesis, this necessitates the use of a large, controllable external magnetic field. This was done using a Newport Instruments iron-core electromagnet, drawing up to 30 amps of current from an auxiliary generator with a maximum output of 10 kW. The current supply was also provided by Newport Instruments, while an additional sweep unit (built in-house) allowed one to change the current supplied to the electromagnet at a fixed rate. The large pole faces provide good field homogeneity and the magnet can safely apply a magnetic field strength of up to 1 Tesla before overheating becomes an issue. The two black cylinders in Fig. 3.17a house the water-cooled coils. The poles are also fitted with modulation coils which can apply an oscillating field of up to 20 Oe and are useful for magnetic resonance experiments. The entire system can be rotated through 360° about the vertical axis and is contained within a fixed rack to which samples and instruments can be mounted. Figure 3.17b shows an image of the sample mount fixed between the poles. Also visible in (b) is the Hall probe used to measure the magnetic field strength. It was built in-house and operates using a simple semiconductor chip. The field dependent voltage is measured by a Hewlett Packard multimeter, HP 34401A. This is then connected to a data acquisition computer (DAC) via a GPIB interface.

Fig. 3.17 Image of Newport Instruments electromagnet used in experiments

3.5.2 Lock-in Amplifier

A lock-in amplifier is an instrument that allows one to detect very small DC signals. It is cleverly designed to lift the small amplitude signal out of the $1/f$ noise through heterodyning. The operational principle is as follows: consider an experiment, whereby the diligent DPhil student wishes to measure how some variable y depends on the value of some controllable parameter x. Now suppose the value of y corresponds to a DC signal that is impossible to discriminate due to $1/f$ noise. The student, by modulating the independent variable x, at a frequency f_{mod}, will cause the value of the dependent variable y, to oscillate at f_{mod} as well. Now that the signal to measure has changed from a small DC signal to a small AC signal, it is time for the student to employ the lock-in amplifier. The output of the experiment will contain an arbitrary sum of signals with different frequencies and phases. Upon input to the lock-in amplifier, the composite signal is multiplied by a sinusoidal wave at a known reference frequency, usually[7] chosen to be f_{mod}. The resultant wave is then

$$\sum_i \left[A_i \cos(2\pi f_i t + \phi_i) \right] \times B \cos(2\pi f_{\text{mod}} t), \tag{3.5}$$

where the sum represents the total input (including noise) to the lock-in amplifier. The multiplication of cosines however, may be rewritten using the sum-and-difference terms, which gives

$$\sum_i \left[\frac{A_i B}{2} \left\{ \cos\left(2\pi(f_i - f_{\text{mod}})t + \phi_i\right) + \cos\left(2\pi(f_i + f_{\text{mod}})t + \phi_i\right) \right\} \right]. \tag{3.6}$$

Inspecting this equation, we note that the small signal of interest, oscillating with frequency $f_i = f_{\text{mod}}$, will now appear to the lock-in amplifier as a DC signal. Furthermore, the $1/f$ noise is now centred around f_{mod}, and with a sufficiently strong low-pass filter may be removed. What remains is an amplified DC signal corresponding to the dependent variable y, having 'locked-in' to its modulation frequency.

The particular lock-in amplifier used throughout this thesis was the Stanford Research Systems model SR850, shown in Fig. 3.18. This model has an input sensitivity range from 2 nV to 1 V, with a dynamic range of over 100 dB. The possible range of modulation frequencies is from 1 mHz up to 102 kHz. Data output from the lock-in amplifier was recorded by the DAC using a GPIB interface.

---

[7]The student could also set the lock-in reference frequency to $3 f_{mod}$ if, for example, the value of x was modulated as with a square wave, where higher harmonic components are also present in the signal corresponding to y.

Fig. 3.18 Image of Standford Research SR850 lock-in amplifier

3.5.3 Spectrum and Network Analyser

An instrument that was vital to the experiments described in Chaps. 4 and 5, was the Rhode & Schwarz ZVL Network Analyzer. This device, shown in Fig. 3.19, also possesses the functionality of a spectrum analyser and allows the user to choose between operating in spectrum analyser mode or network analyser mode. The network analyser allows the measurement of the scattering parameters (S-parameters). These describe the characteristic behaviour of an electrical network by treating it as a 'black box' containing any number of electrical components. The system is characterised by its interaction with the outside world via two ports using what is referred to as the S-parameter matrix, which can be used to predict how the system will respond to signals that are applied to a port. S-parameters are useful for testing how much power is reflected from a port and transmitted to another port. Observing how these values change with respect to an applied magnetic field is a common method for the detection of inductively excited spin waves.

The spectrum analyser mode was used extensively throughout this thesis. Operating in this mode, the user inputs a signal which is then presented as a decomposition of its individual frequency components plotted against the power of said component. The operational principle is similar to that of the lock-in amplifier. The spectrum analyser multiplies the input signal by a reference signal oscillating at f_{ref} and applies a strong low-pass filter, recovering the amplitude of the component of input signal that is oscillating with a frequency f_{ref}. The spectrum analyser repeats this process, scanning between a chosen frequency span with frequency step size also chosen by the user.

The spectrum analyser can also be used in 'zero-span' mode. In this operational mode, the reference frequency is fixed; that is, the span of frequencies measured is zero. The user is then presented with a power-versus-time graph, for a specific frequency component of the input signal, in contrast to the normal mode that shows

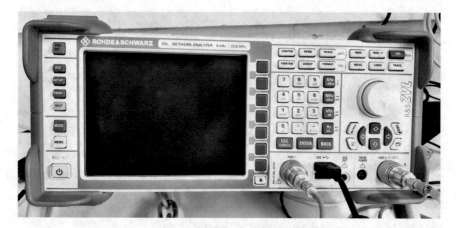

Fig. 3.19 Image of Rhode & Schwarz ZVL network analyzer

power-versus-frequency. This mode is useful for applying the 'lock-in' technique at gigahertz frequencies, as employed in Chap. 4.

3.5.4 Microwave Sources

Multiple microwave frequency sources were utilized for the experiments. In particular, the Hewlett Packard signal generators, 8671A and 8672A were most often used, and are both shown in Fig. 3.20. The source 8671A has a frequency range

Fig. 3.20 Image of microwave sources used in experiments. Synzthesizers are Hewlett Packard 8671A and 8672A

from 2 to 6.2 GHz, and a fixed power output of approximately 18 dBm between 3.9 and 4.1 GHz, the range of interest for this thesis. The other source, HP 8672A can generate signals from 2 to 18 GHz at a range of powers from −30 dBm to 17 dBm within the range of interest. With both sources, the frequency can be tuned to within 1 kHz. Frequency modulation is possible for both of the synthesizers: 8671A allows deviation from the carrier frequency of 0.1 and 10 MHz, while the 8672A allows 0.03, 0.1, 0.3, 1, 3, and 10 MHz deviation from the carrier. In addition to the improved range of frequency modulation, 8672A also offers the option of amplitude modulation at 30 or 100% of the full amplitude.

3.5.5 RF Components

The spin waves that were investigated as part of this thesis possessed frequencies that lay in the radio-microwave frequency region of the spectrum. It was therefore natural to study these spin waves using various RF electronic components. The specific components that were particularly useful (Fig. 3.21) are discussed in Chap. 4 though we take this opportunity to introduce them.

1. **Isolator.** Two isolators were used between the power source and the excitation antenna. Due to the imperfect impedance matching of the 50 Ω transmission line and the coupling antenna, there were high power reflections that could have damaged the equipment. Isolators can be considered to act on AC current as a diode acts on DC current: power can happily flow in one direction, but is met with

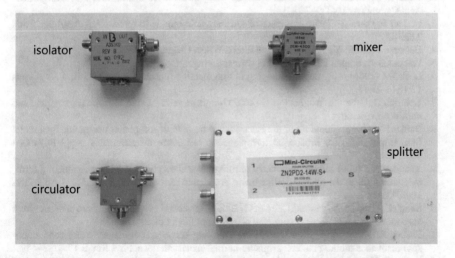

Fig. 3.21 Image of important RF components used. Clockwise from top left: Harris Farinon Ferrites A35369 REV B isolator, Mini-Circuits ZEM-4300 mixer, Mini-Circuits ZN2PD2-14W-S+ splitter, Pasternak PE 8402 circulator

approximately 20 dB of attenuation if passing through the isolator in the opposite direction. The Harris Farinon Ferrites A35369 REV B isolators are rated between 4.7 and 6 GHz, but still offer considerable attenuation at the frequencies of interest.

2. **Mixer.** The microwave mixer was manufactured by Mini-Circuits, model ZEM-4300. The component accepts two inputs which are multiplied together to form the output. It is commonly used for heterodyning and homodyning. The inputs are rated between 0.3 and 4.3 GHz, while the output is rated from DC up to 1 GHz.

3. **Splitter.** The splitter, also manufactured by Mini-Circuits, was used to bisect the power of some input into two outputs. This particular model ZN2PD2-14W-S+ was chosen due to its high power rating, with an operational frequency band 0.5–10.5 GHz fully covering the frequencies of interest. The splitter was used to ensure that the pump antennae were both driven by the same source with the same power.

4. **Circulator.** During the experiments, the probe antenna was used as an excitation and detection antenna. To use the antenna in this way with continous signals, a directional coupler (circulator) had to be used to prevent the excitation source from saturating the detector. The particular circulator used was the Pasternak PE 8402 rated between 4–8 GHz.

References

1. Cherepanov V, Kolokolov I, L'vov V (1993) The saga of YIG: spectra, thermodynamics, inter-action and relaxation of magnons in a complex magnet. Phys Rep 229:81–144
2. Geller S, Gilleo MA (1957) Structure and ferrimagnetism of yttrium and rareearth- iron garnets. Acta Crystallogr 10:239–239
3. Bertaut F, Forrat F (1956) Structure of ferrimagnetic rare-earth ferrites. CR Acad Sci, Paris 242:382
4. Helszajn J (1985) YIG resonators and filters. Wiley, Chichester
5. Prabhakar A, Stancil DD (2009) Spin waves. Springer, Boston
6. Lock JM (1966) Eddy current damping in thin metallic ferromagnetic films. Br J Appl Phys 17:1645–1647
7. Heinrich B, Urban R, Woltersdorf G (2002) Magnetic relaxation in metallic films: single and multilayer structures. J Appl Phys 91:7523
8. Bailleul M (2013) Shielding of the electromagnetic field of a coplanar waveguide by a metal film: implications for broadband ferromagnetic resonance measurements. Appl Phys Lett 103:192405
9. Kostylev M (2009) Strong asymmetry of microwave absorption by bilayer conducting ferro-magnetic films in the microstrip-line based broadband ferromagnetic resonance. J Appl Phys 106:043903
10. Saitoh E, Ueda M, Miyajima H, Tatara G (2006) Conversion of spin current into charge current at room temperature: inverse spin-Hall effect. Appl Phys Lett 88:182509
11. Silsbee RH, Janossy A, Monod P (1979) Coupling between ferromagnetic and conduction-spin-resonance modes at a ferromagnetic-normal-metal interface. Phys Rev B 19:4382–4399
12. Kajiwara Y et al (2010) Transmission of electrical signals by spin-wave interconversion in a magnetic insulator. Nature 464:262–266
13. Kiselev SI et al (2003) Microwave oscillations of a nanomagnet driven by a spinpolarized current. Nature 425:380–383

14. Ando K et al (2008) Electric manipulation of spin relaxation using the spin Hall effect. Phys Rev Lett 101:036601
15. Liu L, Moriyama T, Ralph DC, Buhrman RA (2011) Spin-torque ferromagnetic resonance induced by the spin Hall effect. Phys Rev Lett 106:036601
16. Shiomi Y, Saitoh E (2014) Paramagnetic spin pumping. Phys Rev Lett 113:266602
17. Chumak AV et al (2012) Direct detection of magnon spin transport by the inverse spin Hall effect. Appl Phys Lett 100:082405
18. Sandweg CW et al (2011) Spin pumping by parametrically excited exchange magnons. Phys Rev Lett 106:216601
19. Valenzuela SO, Tinkham M (2006) Direct electronic measurement of the spin Hall effect. Nature 442:176–179
20. Jungfleisch MB et al (2015) Thickness and power dependence of the spin-pumping effect in YIG-Pt heterostructures measured by the inverse spin Hall effect. Phys Rev B 91:134407
21. Wang HL et al (2014) Scaling of spin Hall angle in 3d, 4d, and 5d metals from YIG/Metal spin pumping. Phys Rev Lett 112:197201
22. Schreier M et al (2015) Sign of inverse spin Hall voltages generated by ferromagnetic resonance and temperature gradients in yttrium iron garnet | platinum bilayers. J Phys D Appl Phys 025001:1–5
23. Navabi A et al (2019) Control of spin-wave damping in YIG using spin currents from topological insulators. Phys Rev Appl 11:034046
24. Kryshtal RG, Medved AV (2015) Nonreciprocity of spin waves in magnonic crystals created by surface acoustic waves in structures with yttrium iron garnet. J Magn Magn Mater 395:180–184
25. Kryshtal RG, Medved AV (2012) Surface acoustic wave in yttrium iron garnet as tunable magnonic crystals for sensors and signal processing applications. Appl Phys Lett 100
26. Chowdhury P, Dhagat P, Jander A (2015) Parametric amplification of spin waves using acoustic waves. IEEE Trans Magn 51
27. Weiler M et al (2012) Spin pumping with coherent elastic waves. Phys Rev Lett 108:1–5
28. Weiler M et al (2011) Elastically driven ferromagnetic resonance in nickel thin films. Phys Rev Lett 106:117601
29. Polzikova NI, Alekseev SG, Luzanov VA, Raevskiy AO (2018) Electroacoustic excitation of spin waves and their detection due to the inverse spin Hall effect. Phys Solid State 60:2211–2217
30. Dreher L et al (2012) Surface acoustic wave driven ferromagnetic resonance in nickel thin films: theory and experiment. Phys Rev B 86:134415
31. Fung TC (2015) Phonon magnonics. PhD thesis, University of Oxford
32. Griffiths JHE (1946) Anomalous high-frequency resistance of ferromagnetic metals. Nature 158:670–671
33. Ishak WS (1988) Magnetostatic wave technology: a review. Proc IEEE 76:171–187
34. Zhang Y et al (2018) Antenna design for propagating spin wave spectroscopy in ferromagnetic thin films. J Magn Magn Mater 450:24–28
35. Vasyuchka VI et al (2010) Non-resonant wave front reversal of spin waves used for microwave signal processing. J Phys D Appl Phys 43:325001
36. Karenowska AD, Chumak AV, Serga AA, Gregg JF, Hillebrands B (2011) Employing magnonic crystals to dictate the characteristics of auto-oscillatory spin-wave systems. J Phys Conf Ser 303:012007
37. Demidov VE, Urazhdin S, Demokritov SO (2009) Control of spin-wave phase and wavelength by electric current on the microscopic scale. Appl Phys Lett 95:93–96
38. Karenowska AD, Gregg JF, Chumak AV, Serga AA, Hillebrands B (2011) Spin information transfer and transport in hybrid spinmechatronic structures. J Phys Conf Ser 303:012018
39. Drozdovskii AV, Kalinikos BA, Ustinov AB, Stashkevich A (2016) Spinwave self-modulation instability in a perpendicularly magnetized magnonic crystal. J Phys Conf Ser 769
40. Ganguly A, Webb D (1975) Microstrip excitation of magnetostatic surfacewaves: theory and experiment. IEEE Trans Microw Theory Tech 23:998–1006
41. Demidov VE et al (2009) Excitation of microwaveguide modes by a stripe antenna. Appl Phys Lett 95:10–13

42. Dmytro C (2007) High frequency behaviour of magnetic thin film elements for microelectronics. PhD thesis, Technische Universitat Dresden, pp 29–31
43. Phillips TG, Rosenberg HM (1966) Spin waves in ferromagnets. Rep Prog Phys 29:307
44. Brächer T (2015) Parallel parametric amplification of spin waves in ferromagnetic microstructures. PhD thesis, Technischen Universität Kaiserslautern, pp 63–66
45. ANSYS HFSS: High Frequency Electromagnetic Field Simulation Software (2019) www.ansys.com/Products/Electronics/ANSYS-HFSS
46. Maxwell JC (1862) On physical lines of force. London, Edinburgh, Dublin Philos. Mag. J. Sci. 23:85–95
47. AutoCAD for Mac and Windows (2019) https://www.autodesk.com/products/autocad/overview

Chapter 4
Magnonic Phase Conjugation Experiment

The truth is rarely pure and never simple

Oscar Wilde The Importance of Being Earnest

In this chapter the main results regarding the demonstration of a phase conjugate in a magnonic system are presented. It is suggested through a series of experiments and numerical simulations that the generation of a phase conjugate magnon via FWM is observed.

4.1 Introduction

As discussed earlier, the aim of this experiment was to investigate the possibility of generating a phase conjugated magnon in the continuous wave regime, utilizing a four-wave mixing (FWM) technique. The implementation of the FWM technique relied on the confluence of two pump beams and a probe beam interacting to produce a fourth beam: the phase conjugate of the probe.

4.1.1 General Configuration

The same general configuration was used for all experiments. It comprised a YIG waveguide on GGG substrate, placed YIG-side-down on a PCB designed with three antennae. A sketch of the configuration presented in Fig. 4.1. The illustration shows the pump antennae that create a periodic potential at one end of the waveguide. Perpendicular to these, 3 mm from the nearest point of the pump antennae was the probe antenna that served as a transmitter—sending magnons to 'probe' the pumped region—and as a receiver, collecting the response of the system for analysis. The

© The Editor(s) (if applicable) and The Author(s), under exclusive license
to Springer Nature Switzerland AG 2020
A. Inglis, *Investigating a Phase Conjugate Mirror for Magnon-Based Computing*,
Springer Theses, https://doi.org/10.1007/978-3-030-49745-3_4

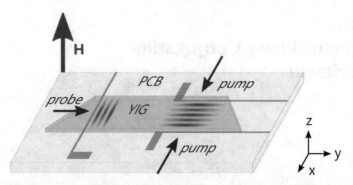

Fig. 4.1 Experimental configuration: yttrium iron garnet (YIG) film with 45° edges placed on PCB with antennae. Counter-propagating pumps excite a standing wave creating a spatio-temporally periodic potential. The probe antenna transmits excitation magnons and receives reflections from the nonlinear region

waveguide was a thin film of width 2.1 mm, length 18 mm and thickness 7.8 mm, with edges cut at 45° to reduce reflections. The waveguide was magnetised perpendicular to the plane, such that only FVMSW are excited. As discussed in Sect. 1.3.3 this field geometry was necessary since the antenna geometry requires magnons propagating orthogonally to one another, therefore requiring the isotropic dispersion offered by FVMSWs.

As discussed in Chap. 2, the excitation of the magnon, m_c, occurs because of the presence of a driving term proportional to m^3, which is the superposition of all oscillating magnetisations at a given region of the waveguide. The phase conjugate magnon will only be excited in a region supporting two counter-propagating pump beams and the probe beam with which it is conjugated. Due to the auto-retracing nature of the phase conjugate magnon, the region where it is excited may be thought of as a phase conjugate mirror (PCM). The spatial separation between the probe antenna and the pumped region ensures that any signal detected by the probe antenna acting as a receiver has been retroreflected.

4.2 The Pumped Region

A large amplitude pumped region comprising counter-propagating waves, is an essential ingredient in the excitation of a phase conjugate wave via four-wave mixing. This is as true in the optical regime as it is the magnonic. Recalling the result from Sect. 2.3.3, that

$$m_c \propto m_1 m_2 m_p^*, \tag{4.1}$$

where m_1 and m_2 are the amplitudes of the pump magnons, it is clear that the greater m_1 and m_2, the greater the amplitude of the phase conjugate signal.

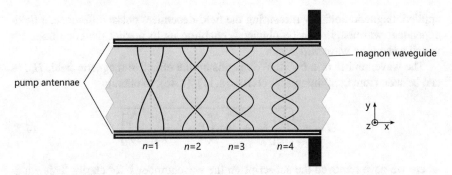

Fig. 4.2 Sketch of standing wave modes for two pump antennae. A resonant mode is supported for spin waves with wavelength $\lambda = 2w/n$ where n is the number of nodes across the width, w

In an optical experiment, the pumped region is formed of a standing wave which can be achieved in two ways: (1) have two counter-propagating waves from identical sources propagating without reflection (2) Use one source propagating normal to a fully reflective mirror. In both of these cases, the set-up is clean and a standing wave is formed at all wavelengths.

The magnonic case is not so simple. Any spin wave excited in our system by a pump antenna is reflected by the opposite pump antenna and waveguide edge. The reflected spin wave is then reflected again from the original antenna with this process continuing, resulting in the confinement of the spin wave to the width of the waveguide. The system, similar to a Fabry–Perot cavity that resonates at specific wavelengths, is illustrated in Fig. 4.2.

For a standing wave to occur across the waveguide width, the magnon wavelength, λ, must satisfy the equation

$$\lambda = \frac{2w}{n} \tag{4.2}$$

where w is the width of the waveguide and n is the order number. If this condition is not met, the phases of the reflected spin waves will be such that the total amplitude within the cavity will sum to zero, and destructive interference takes place. In this case a standing wave will not form and a phase conjugate spin wave will not be generated. It was therefore paramount that a standing wave is excited across the waveguide width. In this section we describe an experiment that investigates how the wavelength of the spin waves depends on the applied field, and how the different spin wave modes fit across the waveguide to excite a standing wave.

4.2.1 Homodyne Experiment

In this experiment the phase difference between a reference signal and a magnon signal propagating across the width of the waveguide was measured as a function of applied magnetic field. By measuring the field-dependent phase difference, a field-dependent wavelength can be obtained, enabling us to predict the conditions for which a standing wave will occur.

The wavenumber of a FVMSW depends on the effective magnetic field, H_0, as can be seen from combining Eqs. (1.10), (1.11), (1.46), to obtain:

$$\omega = \gamma \sqrt{H_0 \left[H_0 + M_{\text{sat}} \left(1 - \frac{1 - e^{-kd}}{kd} \right) \right]}, \qquad (4.3)$$

where we have removed the subscript on the wavenumber k for clarity. This equation describes a powerful phenomenon: holding all else constant, one can tune the wavelength by simply controlling the effective magnetic field. In other words, the refractive index—and therefore the path length—of the magnon waveguide is a function of the applied magnetic field.

Now, consider a spin wave propagating from one pump antenna to the other. At any fixed time, the difference of phase of the wave between these two antennae is given by $\phi(H_0) = k(H_0)w$, where w is the waveguide width, which is constant. Therefore, by measuring the change in phase, $\Delta\phi$, as a function of field, we can extract a change in wavenumber as a function of field, Δk.

The field-dependent phase was measured using the homodyne technique. In this technique, the signal with the field-dependent phase was multiplied by a reference signal of the same frequency. This process returns the superposition of two signals with frequencies equal to the 'sum-and-difference' of the input frequencies.

In the experiment, a microwave signal proportional to $\cos(\omega t)$ was split into two channels with amplitudes A and B. Channel A acted as the reference signal and was fed directly into one input of a mixer. Channel B was used to excite spin waves propagating from one pump antenna to the other, introducing some phase difference $\Delta\phi(H_0)$, before this signal was also fed into the mixer. The mixer multiplies these two inputs together to give an output signal:

$$A\cos(\omega t) \times B\cos\left(\omega t + \Delta\phi(H_0)\right) = \frac{AB}{2}\left[\cos\left(\Delta\phi(H_0)\right) + \cos\left(2\omega t + \Delta\phi(H_0)\right) \right]. \quad (4.4)$$

The RHS of this equation is the sum of a DC term and a fast-oscillating term with frequency 2ω. The latter component may be removed by applying a low-pass filter to obtain the field dependent signal

$$\frac{AB}{2}\cos\left(\Delta\phi(H_0)\right) = \frac{AB}{2}\cos\left(\Delta k(H_0)w\right). \qquad (4.5)$$

Measuring this signal allows us to ascertain the field dependence of the number of complete spin waves across the waveguide, and thus when a standing wave will occur.

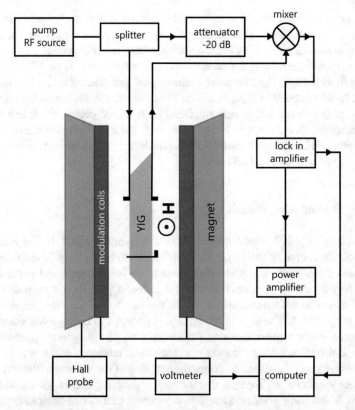

Fig. 4.3 Schematic of the set-up for homodyning experiment. The microwave power is split between an attenuated reference signal, and the magnon excitation signal. The output is homodyned and then passed to a lock-in amplifier. The reference of the lock-in amplifier is itself amplified and used to drive a pair of modulation coils. For a given field, the output of the lock-in amplifier is recorded by a data acquisition computer, as is the corresponding output from the Hall probe

4.2.1.1 Set-Up

Figure 4.3 shows a schematic of the experiment. The output from a microwave power source (Hittite HMC-T2100) was passed into a power splitter. One half of this was attenuated by 20 dB and used as a reference signal, while the other half was used to drive one of the pump antennae. This driving signal excited magnons—with a field dependent wavenumber, k—that propagated a distance w to the other pump antenna for detection. These coupled to the detection antenna to excite an electrical response which was then multiplied with the reference signal using a Mini-Circuits RF mixer. The output of the mixer was a function of the phase difference, ϕ between the reference signal and the magnon signal as described in Eq. (4.4).

The output of the mixer was measured using the lock-in technique. The output of the lock-in amplifier was recorded on a data acquisition computer via a GPIB connection as were the corresponding voltages from the Hall probe.

The output of the microwave source was set to a frequency $f_1 = 3.915$ GHz with a power $P_1 = 0$ dBm. The internal reference signal generator of the lock-in amplifier was set to a frequency of $f_{mod} = 84$ Hz. This reference from the lock-in amplifier was 1 V peak-to-peak which was amplified up to 100 V peak-to-peak and used to drive the modulation coils, which applied an oscillating field of approximately 20 Oe in strength. The low-pass filtering applied to the lock-in amplifier had a time constant of 300 ms and was set to 24 dB/oct.

4.2.1.2 Homodyning Results

The results of the homodyning experiment are found in Fig. 4.4. The blue trace represents the output of the mixer after amplification by the lock-in amplifier. One oscillation corresponds to a phase difference of 2π being introduced to the waveguide. The periodicity of approximately 1 oscillation per 10 Oe shows that as the applied magnetic field increases by 10 Oe, the number of wavelengths across the width changes by 1. The sign of the change is deduced to be negative, since as the field increases for a given frequency the excitation of the system approaches the FMR, and therefore the wavelengths approach the dimensions of the waveguide.

The envelope of the signal describes the coupling of the antenna to the spin waves of various wavenumber. We note that the largest coupling occurs for a field around 3050 Oe, which corresponds to a magnon wavelength of the same order as the antenna width. Then, as the field increases, the wavelength gets larger and the coupling falls off in keeping with the antenna design (Sect. 3.3).

The orange trace in Fig. 4.4 represents the expected curve obtained using a mathematical model. Using Eq. (1.46), the FVMSW dispersion for the field range of interest was calculated. From this, the wavenumber was obtained and multiplied by w, to give a field dependent phase, $\phi(H_{ext})$. This was then multiplied by the envelope of the transmission (obtained experimentally) to give an excellent match between the data and theory.

The good match between data and model is also shown in Fig. 4.5 which shows a Fourier transform of the traces in Fig. 4.4. Since the oscillations occur with respect to field, the x-axis of the Fourier transform represents the number of waves per Oersted. Again, blue and orange represent data and model, respectively. Both traces peak at 0.106 waves per Oe, while the data shows a smaller feature around 0.2 waves per Oe. This smaller feature is most likely a 2nd harmonic of the input to the lock-in amplifier.

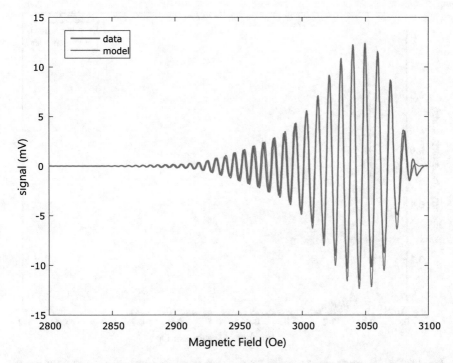

Fig. 4.4 Results of the homodyning experiment. The blue trace represents the data output from the lock-in amplifier as a function of field H_{ext} varying from 2800 to 3100 Oe. The period of the oscillation corresponds to a phase difference of 2π introduced by the YIG waveguide. The orange line shows the expected phase difference that was modelled using Eq. (1.46) and normalised to the experimental data

4.3 Four-Wave Mixing Experiment

Having developed the theory (Sect. 2.3), designed the antennae (Sect. 3.3), and examined the pumped region (Sect. 4.2), we now assemble the puzzle and concentrate our discussion on the four-wave mixing experiment in a magnon waveguide, and the realisation of phase conjugate magnons.

4.3.1 Set-Up

A schematic illustration of the experimental set-up is shown in Fig. 4.6. The pump RF source was the HP 8672A which was set to a power output of 18 dBm and a frequency of $f_1 = f_2 = 3.915$ GHz. The microwave signal was guided in equal amounts into two lines using the microwave splitter and passed through an isolator before reaching the meander antenna. The pump antennae were then used to excite the

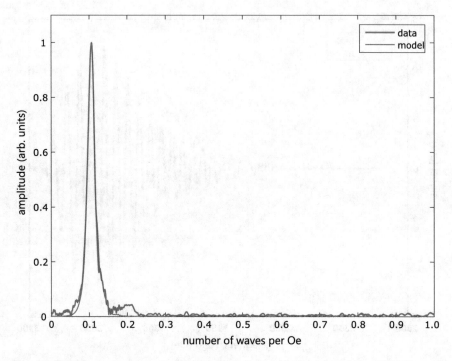

Fig. 4.5 Fourier transforms of traces in Fig. 4.4. Both traces have peaks at a rate of 0.106 waves per Oe. The data (blue) also has a smaller peak around 0.2 waves/Oe which represents the 2nd harmonic of the lock-in amplifier input signal

magnons in the region confined between the pumps. After cable losses and insertion loss, the magnon power was of approximately 40 mW. The probe magnons were excited using the HP 8671A synthesizer, also with a power output of 18 dBm, and a frequency $f_p = 3.91825$ GHz. This signal was passed through a 9 dB attenuator and fed into the probe antenna via the Pasternak circulator. Magnons of approximately 8 mW were excited at the antenna whence they travelled towards the pumped region, suffering propagation loss which reduced the power to approximately 2 mW. Any electrical signals propagating back down the antenna line are then sent to the Rhode & Schwarz ZVL spectrum analyser.

The current supplied to the electromagnet was varied at a rate corresponding to approximately 10 Oe per minute.[1] As the magnetic field was swept the corresponding Hall probe voltage was paired with the appropriate spectrum analyser trace for that field and recorded on the data acquisition computer.

---

[1]Due to the analogue current controller, the sweep-rate of the current was constant, but since the electromagnet was operating in a regime where M is not linear in H, the magnetic field sweep-rate was not constant.

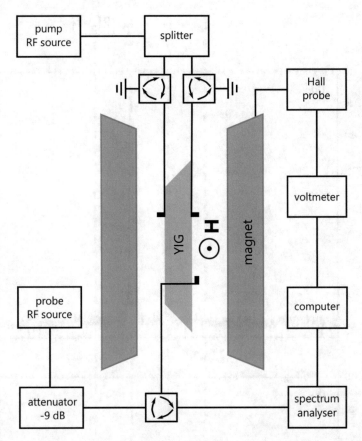

Fig. 4.6 Schematic of the set-up for four-wave mixing experiment. A single microwave source produces a pump signal that is split in half. Each pump line has an isolator between the antenna and the splitter. The probe antenna is driven via a circulator by a separate frequency synthesizer that is attenuated by 9 dB. Any signals returning from the probe antenna are passed to the spectrum analyser and recorded by the data acquisition computer. As the field is swept, the Hall probe sends a corresponding voltage to the voltmeter which is also recorded

4.3.2 Results

A comparison of two spectra for different field configurations can be found in Fig. 4.7. In (a), despite there being zero applied magnetic field the spectrum analyser records two peaks. The peak at $f_1 = 3.915$ GHz is the crosstalk between the pump and probe antennae: the probe antenna is coupling directly to the electrical signals emitted by the pump antennae, which are detected by the spectrum analyser at a power of $P_1 = -42.5$ dBm. The larger peak at $f_p = 3.91825$ GHz is purely electrical. The probe signal travels through the circulator towards the probe antenna, whereupon a large amount is reflected. This reflection then propagates back towards the circulator

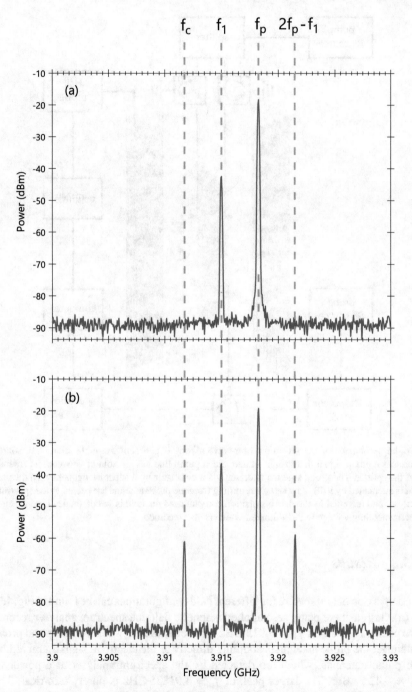

Fig. 4.7 Spectra measured in four-wave mixing experiment. The spectrum in **a** is measured with no magnetic field applied. Both of the peaks are artefacts of the microwave electronics used in the experiment. As a field is increased to 3077 Oe (**b**) two side peaks appear

which couples the reflected signal to the spectrum analyser and measures a signal of approximately $P_p = -18.3$ dBm. As the magnetic field was increased to $H_{ext} = 3077$ Oe the spectrum in Fig. 4.7b was observed. Of particular note is the signal at the expected phase conjugate frequency $f_c = 2f_1 - f_p$ with power $P_c = -61.1$ dBm. The presence of this peak is not inconsistent with the notion that under the correct biasing conditions, phase conjugate reflections may occur from the pumped region. Also present in this spectrum is the unexpected rightmost peak, occurring at $f = 2f_p - f_1$ and power $P = -58.9$ dBm. This peak is likely arising due to a FWM process involving reflections from the waveguide edge farthest from the antenna.

We also note the difference in power of the peaks f_1 and f_p as the magnetic field is applied. The f_1 signal has increased in power from $P_1 = -42.5$ dBm to $P_1 = -36.5$ dBm. This increase is caused by the leakage of pump magnons travelling down the waveguide and being detected by the probe antenna. The f_p signal however, decreased in power from $P_p = -18.3$ dBm to $P_p = -19.2$ dBm. The reduction in power occurs because as the magnon passband coincides with the input frequency, some power is absorbed as a spin wave excitation.

To help characterise the system response, spectra were measured for a full magnetic field sweep. The results of this are shown in Fig. 4.8 which shows a 3D plot of the power detected by the probe antenna versus frequency and applied field. The two solid, horizontal red lines represent the ever-present pump signal and probe reflection. The power of these signals varies little over the recorded field range. Highlighted at 3077 Oe is a yellow vertical dotted line which shows where the spectrum in Fig. 4.7b lies in the field sweep.

Also visible are extra side bands beyond those appearing in Fig. 4.7b. One of these signals appears at $f = 3.9085$ GHz $= 2f_c - f_1 = 3f_1 - 2f_p$, suggesting that there are higher order mixing processes occurring, most likely due to reflections from the edge closest to the pumps.

The boxed region highlights a field range of the phase conjugate signal at f_c, which is inspected in further detail in Fig. 4.9. It is clear that this signal, and the one occurring at $f = 2f_p - f_1$, have some structure to be examined further. Upon examination of the P_c signal with respect to magnetic field, two main features catch the eye.

The first of these is the fast oscillation occurring approximately every 5 Oe. This feature may be neatly explained with reference to Figs. 4.4 and 4.5. Both of these figures tell us that in the applied field region of $H = 3050 \rightarrow 3100$ Oe, for a spin wave frequency of $f_1 = 3.915$ GHz, an increase in field of 10 Oe corresponds to the subtraction of one wavelength across the width of the waveguide. The oscillation period of 5 Oe in Fig. 4.9, suggests that as the number of half-wavelengths across the pumped region changes by one, the efficiency of exciting the signal at f_c undergoes a full oscillation. This result suggests two things:

- The signal at f_c is strongly dependent on pump amplitude.
- A standing wave is set-up in the pumped region, where supported modes have wavelength $\lambda = 2w/n$, where n is a natural number.

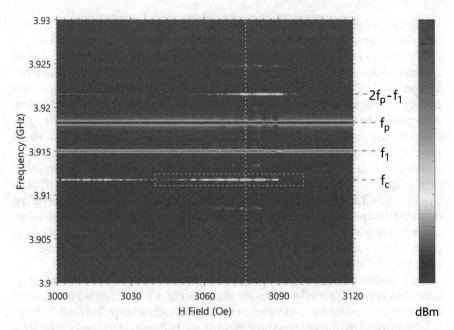

Fig. 4.8 Collection of spectra for applied magnetic fields between 3000 and 3120 Oe. The vertical dotted line at 3077 Oe corresponds to the trace shown in Fig. 4.7b and with the frequencies of interest are highlighted. The orange boxed region emphasises the phase conjugate signal at f_c. The variation of this signal with field is shown in detail in Fig. 4.9

Figure 4.9, depicts a large power variation with field. For example, the signal power decreases from -60.7 dBm at 3077 Oe to -74.7 dBm at 3080 Oe, before rising back up to -60.8 dBm, corresponding to a factor of 25 difference in power. Assuming this oscillation arises from resonant pump modes, this large dependence on the pump amplitude is consistent with the equation for the phase conjugate term derived in Sect. 2.3.3.2, where Eq. (2.46) states the dependence is $m_1 m_2$.

The standing wave modes appear to be strongest when the pump magnons have an integer number of half wavelengths across the waveguide width. In this sense, we may consider the action of each pump antenna on the pumped region as similar to that of a resonant Fabry–Perot cavity.

We also note that within the range $3060 \to 3080$ Oe, the local minima are still of significant amplitude, suggesting that the destructive interference across the pumped region is not perfect. This can be explained by considering the effects of spin wave damping.

Let us consider a wavelength for which the cavity is non-resonant. Figure 4.10 depicts a non-integer number of half-wavelengths across the width of the waveguide, where one antenna is exciting and the other is passive. For clarity, but without loss of generality, we have chosen a wavelength equal to 5 quarter-wavelengths. For the case of the non-damped system, the incident wave and the 1st reflection from the

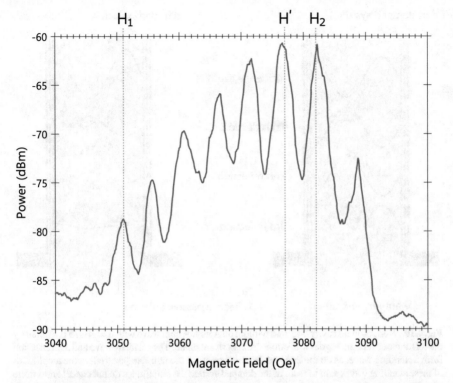

Fig. 4.9 Magnetic field dependence of the signal at $f_c = 3.91175$ GHz recorded by the probe antenna. The field $H' = 3077$ Oe is the same as the spectrum in Fig. 4.7b. The fast oscillation occurring with a period of approximately 5 Oe is caused by the standing wave modes of the pumps. The larger envelope is due to width modes of the probe magnons which are simulated for the fields $H_1 = 3051$ Oe and $H_2 = 3082$ Oe and compared in Fig. 4.15

passive antenna form a standing wave. The 1st reflected wave is then itself reflected from the excitation antenna, causing a 2nd reflected wave. This wave then causes a 3rd reflection and forms a separate standing wave that is π radians out of phase with the first standing wave, resulting in complete destructive interference. It can also be seen that the two blue waves cancel each other perfectly, as do the two red waves.

The picture is not as neat for a damped system, as is the case for spin waves in YIG. We observe that the equivalent standing waves in the damped system, while π radians out of phase, have different amplitudes due to the exponential amplitude decay caused by damping. Therefore in the lossy system there will always be some small but finite standing wave which drives the FWM effect, as observed in Fig. 4.9.

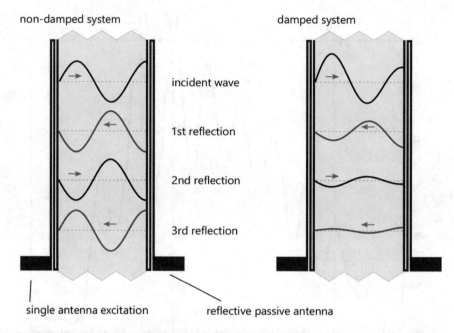

Fig. 4.10 Non resonant cavity with and without damping. A spin wave with $\lambda = 4w/5$ is excited by the left antenna and undergoes reflections from both antennae. The incident wave and 1st reflection form a standing wave, as do the 2nd and 3rd reflections. In the non-damped system the amplitudes of these standing waves sum to zero. In the damped system the amplitudes do not cancel and a finite standing wave exists

4.4 But Is It a Phase Conjugate?

The results presented above demonstrate that there is certainly some FWM process occurring, that generates the signal with frequency f_c. The following section presents further evidence to help elucidate the nature of this effect, and to develop further a physical picture of this process.

4.4.1 Time Lag Measurements

In trying to test the existence of a phase conjugate mirror, a well defined pumped region that is sufficiently far away from the probe antenna is essential. By spatially separating the transmission/detection region from the interaction region, it allowed for the determination of whether the nonlinear signals were reflections from the pumped region or whether they were a nonlinear effect local to the probe antenna.

A similar experiment to this was performed using spin wave bullets [1]. A pulse was sent down a waveguide before interaction with a parametrically pumped region.

The original pulse, on its way to the pumped region suffered some dispersion, which upon reflection from the region was undone. This process was measured directly, using the imaging technique Brillouin light scattering (BLS) unavailable in the forward volume configuration.

In the absence of BLS, pulsing the excitation and measuring the system response time was a method at our disposal. Pulsed measurements however, offered their own challenges in the form of RF leakage through the switch, and the spread of spectral components associated with a given pulse. The following experiment describes a method where these issues were avoided and a time lag between excitation and reflection was measured.

4.4.1.1 Set-Up

The experimental set-up is shown in Fig. 4.11. The experimental configuration is similar to that for FWM, but with two crucial differences:

- an arbitrary waveform generator (AWG) is introduced as a trigger source.
- the spectrum analyser is used in zero-span mode.

The arbitrary waveform generator is included as a trigger source for the spectrum analyser and the probe power source. The addition of the AWG allowed the probe RF source to be driven with amplitude modulation, such that if a signal is reflected from the pumped region and depends linearly on the amplitude of probe spin waves, the reflection must also have the same amplitude modulation profile. Furthermore, by running in the spectrum analyser in zero span mode, we can directly observe the amplitude modulation for a given carrier frequency.

The time delay of the phase conjugate reflection is measured by comparing the phase difference of the amplitude modulation of two signals: the original probe signal at f_p and the signal appearing at the phase conjugate frequency f_c. The impedance mismatch of the probe antenna allowed us to use the original probe signal at f_p as a reference signal since such a large proportion of the power is reflected from the antenna towards the spectrum analyser. Furthermore, since the driver for the modulation of this reference signal is the same as the trigger for the spectrum analyser, the reference is very stable. This reference signal is obtained using the spectrum analyser in zero span mode and specifically tuning to f_p.

The spectrum analyser is then used to measure any response of the system at the frequency f_c. Any excited signal at this frequency, that has amplitude proportional to the probe amplitude, will necessarily have the same amplitude modulation. Moreover, the modulation of the f_c signal will be out of phase with the reference signal if it has originated at some point that is not local to the probe antenna. The phase difference will correspond to the time taken for the probe magnons to travel from the antenna to the interaction region, summed with the time taken for the generated f_c signal to return from that region.

In the experiment, the pump source was kept at $f_1 = f_2 = 3.915\,\text{GHz}$, with the probe frequency at $f_p = 3.91825\,\text{GHz}$. The field was swept over the same region

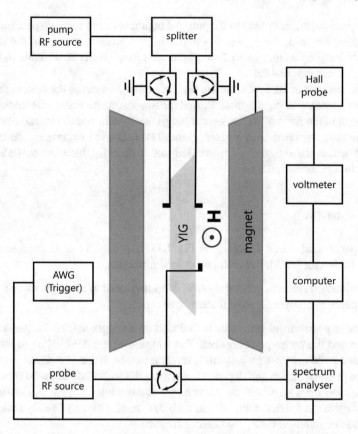

Fig. 4.11 Experimental set-up of time lag measurement. The inclusion of the arbitrary waveform generator (AWG) allows the probe source to be amplitude modulated at a given frequency, while also triggering the spectrum analyser trace set to zero span mode

of interest 2960 Oe $\rightarrow$ 3100 Oe. The AWG was used to generate a sine wave of 5 V peak-to-peak, to act as the modulation and trigger source for the range of frequencies $f_{AM} = 400, 200,$ and 100 kHz. The AWG output was split using a BNC tee-piece, with an attenuator on the line that was sent to the microwave source since the maximum input signal for the modulation source was 1 V peak-to-peak.

For each different modulation frequency, two field sweeps were performed: one with the spectrum analyser tuned to the reference signal f_p, and the other to obtain the nonlinear signal at f_c. For each modulation, the reflected signal was recorded at the same field ($H = 3015$ Oe), indicated by the dashed in line Fig. 4.12a, which shows the time-averaged response of the system for signals of frequency f_c.

Figure 4.12b–d show the time-dependent responses for the different values of f_{AM}, the modulation frequency. Each measurement was taken over 4 periods of modulation. In each of the traces, there is a phase lag (and corresponding time lag) between the reference signal (black) and the response signal (red). For each different

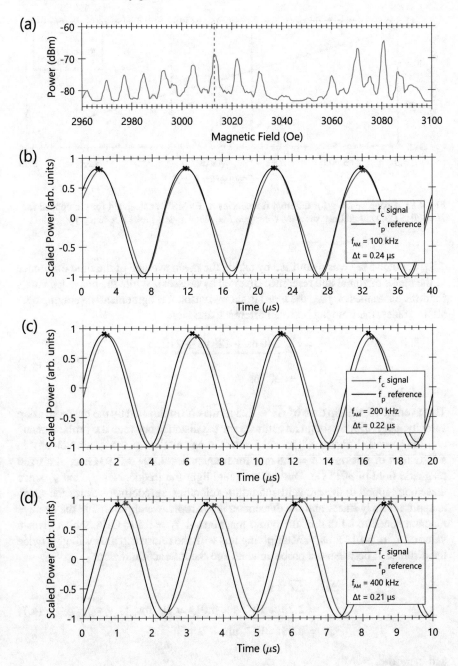

Fig. 4.12 **a** Time-averaged zero-span measurement of system response at f_c. The dashed line shows the applied field for which traces (**b**)–(**d**) were obtained. The time dependent responses for different values of f_{AM} are given in (**b**)–(**d**) over 4 periods of modulation. The phase differences between the response (red) and the reference (black) are observed by comparing the crosses. For various modulation frequencies, the phase lag is different, but the time lag, Δt, is similar

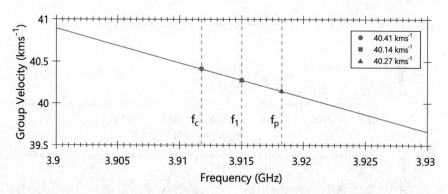

Fig. 4.13 Group velocity for different frequencies of FVMSWs calculated for the applied field $H = 3015$ Oe. Dashed lines show the conjugate frequency f_c and probe frequency f_p

f_{AM}, the phase lag was quantified by taking the mean average of the time difference of the peaks of signal and reference (marked by crosses). While the *phase* lag varies for different values of f_{AM}, the *time* lag shows reasonable agreement. Averaging over all the traces, the time lag between reference and signal is

$$\overline{\Delta t} = \frac{240\,\text{ns} + 220\,\text{ns} + 210\,\text{ns}}{3}$$

$$= 223\,\text{ns}. \tag{4.6}$$

This average return trip time of $\overline{\Delta t} = 223$ ns may be multiplied by the magnon group velocity, $v_g(f, H)$, to obtain an estimate for the distance between the probe antenna and the so-called mirror. Figure 4.13 shows a calculation of the group velocity in a YIG film of thickness $d = 7.8$ mm for frequencies $3.90 \rightarrow 3.93$ GHz, at a fixed magnetic field of 3015 Oe. Dashed lines highlight the frequencies f_p and f_c since these correspond to the outward and return velocities, respectively. Since the graph is approximately linear and the dispersion is isotropic, we approximate the average magnon speed to be that of the pump magnons at $f_1 = 3.915$ GHz, which gives a value of $v_g = 40.14$ kms^{-1}. Multiplying this with the return trip time will give twice the distance L between the probe antenna and the interaction region:

$$2L = \overline{\Delta t} \times v_g$$

$$= 2.23 \times 10^{-7}\,\text{s} \times 4.014 \times 10^4\,\text{ms}^{-1} \tag{4.7}$$

$$= 8.98 \times 10^{-3}\,\text{m},$$

and therefore,

$$L = 4.49 \times 10^{-3}\,\text{m}. \tag{4.8}$$

This result of 4.49 mm is to be considered in context of the designed antennae on the PCB. The spacing between the probe antenna and the nearest edge of pump antennae is 3 mm, which then extend another 4 mm farther. The obtained value for L, is therefore well within the expected range of distances from which a reflection may occur. This result is not inconsistent with the assertion that the signal at f_c is indeed a reflection from a pumped nonlinear region.

4.4.2 Simulations

As a method of offering further insight into the FWM experiment, simulations were performed. The calculations were implemented in MuMax3 [2], a software package specialising in micromagnetic simulation. As with all computer simulations of physical systems, the user must make certain concessions and omissions when transitioning from the real to the virtual system.[2] Due to limited computational resources, a full-scale exact replica of the experiment was not tenable, though a modified system that retained the central aspects of the experiment was designed. The aim of the simulation was to aid the explanation of the experimental results, specifically:

- To explain the presence of the peak at $f = 2f_p - f_1$.
- To confirm that the f_c signal is a reflection from the pumped region.

The system that was replicated and studied is depicted in Fig. 4.14. The simulated universe had dimensions of 8 mm $\times$ 3.1 mm $\times$ 7.8 mm, with the total number of cells equal to 2048 $\times$ 512 $\times$ 1. A region (with dimensions 7 mm $\times$ 2.1 mm $\times$ 7.8 mm) with the physical parameters of YIG was created to act as the waveguide. The chosen parameters were saturation magnetisation $M_{sat} = 138.6 \, \text{kAm}^{-1}$, exchange stiffness, $A_{ex} = 3.5^{-12} \, \text{Jm}^{-1}$, and Gilbert damping, $\alpha = 5 \times 10^{-5}$ [3]. The probe antenna was placed 100 mm from the left edge and extended the full width of the waveguide. The pump antennae were placed at the top and bottom edges of the waveguide, extending 4 mm from the right edge. A detection region was defined 1 mm from the pump antennae, and was 310 mm wide.

To control for the effects of reflections from waveguide edges (rather than reflections from periodic potential in the pumped region) a 'dead zone' was defined at certain waveguide boundaries. These regions were defined to have a damping parameter 300 times greater than the waveguide, therefore allowing magnons to propagate into this region, but diminishing them to a negligible level before they reflect from the universe boundary. By suppressing these edge-reflections, any signals detected at the output may be attributed solely to the nonlinear interaction of the probe magnons with the pumped region.

As with the experiment, the pump antennae were driven at 3.915 GHz and the probe was driven at 3.91825 GHz. At each specific field configuration, the system

---

[2]For example, no physical experiment can be isolated to the same levels of perfection as a simulation.

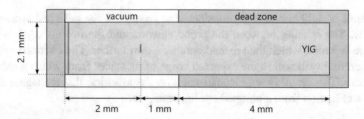

Fig. 4.14 Geometry of the simulated system. The pink region was assigned the parameters of YIG. The grey represents a 'dead-zone' with increased damping to suppress reflections. The red lines represent the antennae with the probe on the left and the pumps on the right. The green line shows the defined detection region

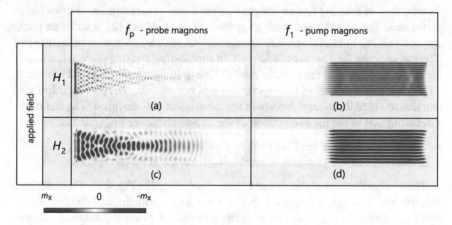

Fig. 4.15 Examples of simulated transverse magnetisation m_x. The rows correspond to different applied magnetic fields of $H_1 = 3051$ Oe and $H_2 = 3082$ Oe. The columns show the isolated behaviour of different input frequencies, $f_p = 3.91825$ GHz and $f_1 = 3.915$ GHz. Subfigures **a** and **c** compare propagation efficiency of the probe magnons, while **b** and **d** show the standing wave created by the pump antennae

was simulated for a window of 2 ms before a Fourier transform was performed on the transverse magnetisation measured at the detector.

Figure 4.15 shows amplitudes of pump and probe magnons for different magnetic fields. The white rectangles depict the simulated waveguides (not the simulated universe), with the amplitude of the x-component of the transverse magnetisation, m_x, represented by the red and blue colouring. The figure allows for a useful qualitative comparison of the behaviour of the pumps and the probe magnons for two different external magnetic fields. These simulations can explain the features observed in the FWM experiment, specifically Fig. 4.9.

Turning our attention to Fig. 4.15a, c, we are able to describe the general envelope of Fig. 4.9. Note that at every point along the waveguide, the probe spin waves excited at $H_1 = 3051$ Oe have a lower amplitude than those excited at the larger field $H_2 = 3082$ Oe. Comparing (a) and (c), it is clear that the lower field corresponds to the appearance of many more nodes across the waveguide width. This demonstrates that when $H_1 = 3051$ Oe a higher width mode is excited with reduced efficiency of propagation. The reduced magnon amplitude evident in (a) is manifested in Fig. 4.9, where a smaller signal is observed at H_1 compared to H_2.

The simulations also offered an insight into the pump behaviour. Figure 4.15b, d show the amplitude of the pump magnons for the applied magnetic fields H_1 and H_2. In both of these a standing wave is clearly visible, with good confinement and uniformity along the length of the antennae. At the lower field, H_1 a wave with 18 nodes across the width is excited, while as the field is increased to H_2, the number of nodes decreases to 12. As the field is increased from H_1 to H_2, the intensity of the standing wave will oscillate with every node that is removed. Therefore, according to our physical model, it is expected that between the two applied fields, the amplitude of the phase conjugate reflection will oscillate $18 - 12 = 6$ times. This behaviour was observed in Fig. 4.9, where the amplitude goes through six oscillations between the fields H_1 and H_2.

A spectrum of the system response was also calculated. After simulating the system for $H = 3077$ Oe for a 2 ms window, a Fourier transform was performed on the time dependent value of m_x at the detector region. The calculated spectrum, shown in Fig. 4.16, has many features worth discussion. The most prominent of these is the large and wide peak at the probe frequency, f_p. The large relative amplitude is due to the placement of the detector: in order for probe magnons to reach the pumped region, they must pass through the detection region first, hence the large response at f_p.

The width of this peak may also be explained by noting that the simulated spectrum was obtained using a Fourier transform of a 2 ms time interval. The raw output therefore comprised a set of theoretical peaks, each of which was convoluted with a sinc function corresponding to the 2 ms window. This sinc function then needed to be filtered out and the vertical axis converted to a log scale for comparison with measurement: this visual processing had the side-effect of broadening the central peak. This also explains why the peaks are neither Gaussian nor Lorentzian.

Also of note is the relatively small power of the peak occurring at f_1. In the experimental results (Fig. 4.7b) this peak is approximately 20 dB larger than the signal at f_c, which is in contrast to the calculation where P_c is greater by 1.6 dB. The difference in P_1 is for two reasons: (1) the simulation eliminates direct coupling contributions (2) the confinement of the magnons between the pump antennae is better in the simulation where reflections are negligible.

The most significant feature of Fig. 4.16 is the peak at f_c. Because the simulated system eliminates reflections from the ends of the waveguide, this signal must necessarily be reflected from the pumped region of periodic potential. Given the direction of travel, and its generation from a FWM process, it must also possess the wavevector $\mathbf{k}_c = -\mathbf{k}_p$. This is one of the key aspects of a phase conjugate reflection and it has been satisfied.

Another feature of the simulated spectrum worth commenting on is the absence of the peak at $2f_p - f_1$. In the experimental data (see Fig. 4.7) this peak of a significant amplitude, comparable with P_c, yet in the simulations the peak has disappeared. The

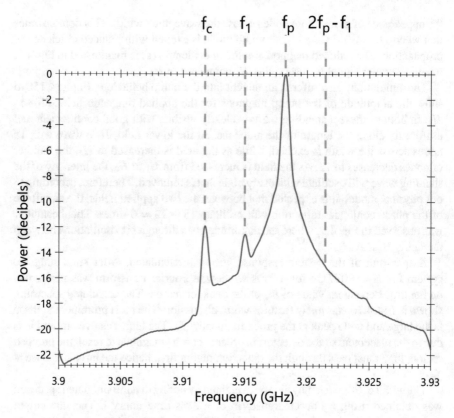

Fig. 4.16 Spectrum from simulated FWM experiment. Applied field is $H = 3077$ Oe (as in Fig. 4.7). The large width of the peak at f_p is due to the limited computation resolution and conversion from linear to logarithmic scale. The peak at f_1 is relatively small since there is no direct coupling. Significantly, a large peak is present at f_c the phase conjugate frequency

removal of the peak is explained however by the suppression of reflections from the back wall of the waveguide.

4.4.3 Non-degenerate Effect

If we briefly remove ourselves from the tight focus of the experiments described above and take a moment to consider a physical concept at the centre of this thesis—four-wave mixing—an obvious question arises: why choose *non-degenerate* FWM, rather than *degenerate* FWM, given the physics is simpler in the latter case?

In this section, we shall endeavour to answer this, and offer a brief description of how the non-degeneracy affects the resultant signal. In this section we address the reasons for choosing the non-degenerate set-up, and the effects of this, theoretically and physically.

4.4.3.1 Degenerate Difficulty

The reasons for avoiding the degenerate set-up were exclusively due to experimental limitations. Our aim of observing a phase conjugate spin wave was reliant on the observation of a reflection of a probe magnon from a pumped region. Given the geometry of our waveguide and the dispersive nature of CW magnons compared to well-confined laser beams, the conventional optical method of discerning a phase conjugate—whereby the probe is incident on the pumped region at some angle that is *not* perpendicular to both the pump beams, therefore destroying the symmetry between traditional reflection and phase conjugate reflection—was not feasible. We therefore had to be wary of regular reflections from air-YIG boundaries, and have a method of discerning them from the signal of interest.

The auto-retracing nature of phase conjugate reflections imposed the further condition that the same probe antenna had to be used for excitation and detection. Under this constraint, it would in principle have been possible to pulse the probe excitation and wait for a response. When attempting this however, with the condition that $f_p = f_1 = f_c$, any delayed f_c response was drowned out by direct coupling to the CW pump signal and microwave leakage through the switch driving the probe pulse that was reflected from the probe antenna. The latter effect could have been overcome by improving the switch isolation and reducing the antenna mismatch, however the direct coupling between the pump and probe is much more challenging.

A solution to this problem was to move to a non-degenerate paradigm. In this regime, conservation of energy ensures that a method of discerning input from output is inherent to the FWM process. By breaking the symmetry between pump frequency and probe frequency, it must also be broken with respect to the phase conjugate frequency. This then allows the observation of a phase conjugate signal using a spectrum analyser, since it is the only signal allowed with frequency $f_c = 2f_1 - f_p$, as discussed in Sect. 2.3.3.

4.4.3.2 Detuning Effect

While a comprehensive theory of magnon phase conjugation via FWM remains to be established, the groundwork for degenerate FWM has been laid in Sect. 2.3. Since this degenerate case was arrived at using, in part, analogous analysis to the optical theorists Yariv and Pepper, intuition suggests that the non-degenerate optical case may offer some clue as to what to expect in the magnetic case.

In general, the non-degenerate case may be characterised by the rule of thumb that the larger the detuning between pump and probe, the less efficient the phase conjugate mirror. It is this property that allows the use of phase conjugate mirrors to be used as optical bandpass filters [4].

Indeed, a similar reduction of efficiency with increasing detuning was observed for the magnonic case as well. An experiment with various detuning levels was performed in an effort to better understand the effect of non-degenerate pump and probe magnons. Using the same set-up as depicted in Fig. 4.6, a series of magnetic field sweeps were performed for different combinations of pump frequency and probe frequency. The frequencies f_1 and f_p were both varied between 3.878 GHz → 3.958 GHz, such that $\Delta f \equiv f_p - f_1$ varied in steps of 8 MHz, with the maximum detuning therefore $\Delta f_{max} = 80$ MHz. For each frequency combination the external magnetic field was swept and a series of spectra were recorded. From these, the maximum power recorded at the frequency $f_c = 2f_1 - f_p$ was extracted and plotted in the matrix shown in Fig. 4.17.

Measurements of the array diagonal, representing the degenerate case where $f_1 = f_p$ were omitted. For the reasons outlined above, degenerate measurements are at best unenlightening and at worst misleading and have been set to an arbitrary value of -50 dBm and shaded black to aid with presentation. The results show the general trend that as one moves farther from the degenerate case, the phase conjugate power, P_c decreases. This phenomenon is in line with the optical regime, where the phase mismatch of Δk manifests itself experimentally as a reduction of reflectivity of the phase conjugate mirror [5].

It should also be noted that the reduction in efficiency with increased detuning is not monotonic, and displays some texture. Following a line from the black, top-right square in Fig. 4.17b, down to the bottom left square we may examine the response from almost degenerate to the maximal detuning of $|\Delta f| = 80$ MHz. This line shows that after hitting the noise floor, the signal picks back up again around $|\Delta f| = 72$ MHz. A similar effect may be observed tracing from the black, bottom-left square up to the top-left square.

Another interesting feature of Fig. 4.17 is the non-reciprocity of the efficiency with respect to detuning. That is, P_c depends not only on the magnitude of Δf, but also the sign. This is evident in (a) where it can be seen that for positive Δf, where $f_p > f_1$, the roll-off of P_c is less steep than for $f_p < f_1$. Furthermore, the amplitude of P_c for positive values of Δf never reaches the noise-floor, in contrast to some values for negative Δf.

The texture and asymmetry of these data may be explained by recalling the antenna reflection efficiencies recorded in Fig. 3.16. Since the antennae are not flat over the region of detuning, the fraction of the electrical power that is converted into spin wave power will vary accordingly. Considering this added variation in input power can help describe the non-reciprocity of the phase conjugate amplitude displayed in Fig. 4.17.

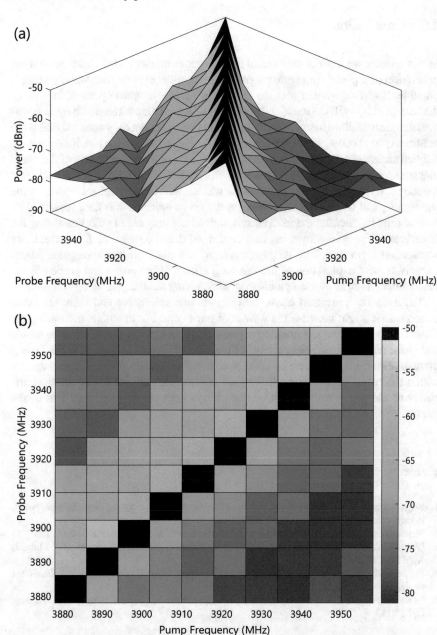

Fig. 4.17 Maximum power of phase conjugate reflection for various pump and probe frequencies, f_1 and f_p, respectively. Both frequencies were varied between 3878 MHz and 3958 MHz. As the external magnetic field was swept, a series of spectra were recorded. The maximum output power of the signal occurring at $f_c = 2f_1 - f_p$ is shown at an elevation (**a**) and from above (**b**). Degenerate measurements where $f_1 = f_p$ were omitted and replaced with an arbitrary value of -50 dBm for the purposes of presentation

4.5 Conclusion

In this chapter we have demonstrated through experiments and simulations that the generation of a phase conjugate magnon from non-degenerate four-wave mixing is possible. It was shown that over the range at which the antenna operates, increasing the bias field by $\approx 10\,\mathrm{Oe}$ reduces the number of waves across the width by 1. It was then demonstrated with experiments and simulations that not only does a signal at the predicted phase conjugate frequency $f_c = 2f_1 - f_p$ exist, but the amplitude of such a signal varies with magnetic field in the same way that a resonant mode of the pump magnons varies; a dependence consistent with the notion that f_c is a phase conjugate reflection. Evidence that the signal at f_c was excited in the pumped region of the waveguide, and was not an effect local to the probe antenna, was then presented by way of time lag measurements. The simulations not only aided with visualising the behaviour of the spin waves, but also confirmed that the signal at f_c is necessarily a reflection from the phase conjugate mirror, and not from the waveguide edges. Latterly, a brief exploration of the detuning effects of the pump and probe offered yet another example of a magnonic system behaving similarly to an optical one.

The evidence presented above however, is not exhaustive and there are other methods that might increase the weight of our claim. For example, measuring the phase conjugate power as a function of pump and probe power should show quadratic and linear relationships, respectively, if the model of Chap. 2 is correct. While testing these relationships, care must be taken to ensure the experimental conditions remain within the range for which the theoretical assumptions are valid. For instance, the power of the probe magnons should be always kept much lower than that of the pumps.

References

1. Serga AA et al (2005) Parametric generation of forward and phase-conjugated spin-wave bullets in magnetic films. Phys Rev Lett 94:167202
2. Vansteenkiste A et al (2014) The design and verification of MuMax3. AIP Adv 4:107133
3. Chumak AV, Serga AA, Hillebrands B (2017) Magnonic crystals for data processing. J Phys D Appl Phys 50:244001
4. Pepper DM, Abrams RL (1978) Narrow optical bandpass filter via nearly degenerate four-wave mixing. Opt Lett 3:212
5. Nilsen J, Yariv A (1979) Nearly degenerate four-wave mixing applied to optical filters. Appl Opt 18:143

Chapter 5
Investigating Nonlinear Effects

L'imagination se lassera plutot de concevoir que la nature de fournir [Imagination tires before Nature does]

Blaise Pascal [1]

As discussed in Chap. 4, the nonlinearity intrinsic to magnonic systems leads to important physical effects for potential device applications. In addition to four-wave mixing, there is a multitude of other nonlinear effects that have been demonstrated in YIG. In this chapter, we report on the emergence of a time-domain signal with a fractal-like structure, observed in the presence of a spatio-temporally periodic potential within a spin wave system. The results are presented and discussed in the context of wider fractal observations in the magnonic community.

5.1 Introduction

The rich variety of nonlinear effects that have been demonstrated in magnonic systems—YIG thin-films in particular—have led to a collection of potential devices. In addition to the phase conjugate mirror demonstrated in Chap. 4, researchers in the magnonic community have developed delay lines [2], soliton train generators [3], and bistable resonators [4] with possible applications for switching and signal processing. The development of nonlinear magnonics is imperative to the development of magnonic computing and novel devices.

A. Inglis, *Investigating a Phase Conjugate Mirror for Magnon-Based Computing*,
Springer Theses, https://doi.org/10.1007/978-3-030-49745-3_5

5.1.1 Solitons

A soliton, or a solitary wave, is a nonlinear phenomenon that may be considered to be a localised wave propagating in one space direction only, with a constant shape that is left unaltered after collision with other solitons [5]. Existing in nonlinear, dispersive media they may be understood qualitatively as embodying a perfect balance between the effect of dispersion on a pulse, and the effect of nonlinearity, such that (excluding loss mechanisms) the unaltered shape will propagate indefinitely.

The discovery of the soliton dates back to 1834 when it was first observed as a novel water wave.[1] Solitons have been observed in magnonic systems since the 1980s, whereby the archetypal soliton shape is an envelope applied to a carrier frequency. This envelope can form either bright or dark solitons, depending on two controlling factors: (1) the dispersion parameter characteristic of the medium, D (2) the nonlinear frequency response parameter, N. When these parameters have opposite sign, the Lighthill criterion is satisfied and attractive nonlinearity allows the formation of bright solitons, while systems with same sign leads to repulsive nonlinearity and dark solitons.

Soliton behaviour in spin wave systems is initiated in two ways. The most straightforward way to excite single solitons is by simply applying appropriate pulses to the system input [7]. The other second method is by introducing a large amplitude CW signal. This large signal forms a train of solitons through a process of modulational instability (MI) [8], wherein the large signal decays through a FWM process to form a series of sidebands around a carrier, resulting in an amplitude modulated signal resembling a soliton train. When this process is initiated by a high powered monochromatic input, it is called 'self-modulation instability' (SMI), though the effect may also be stimulated, in which case the term 'induced modulation instability' (IMI) is used.

Bright envelope spin wave solitons excited from a SMI process have been studied since the 1980s [9]. More recently however, Wu et al. [10] demonstrated the viability of SMI as a method for generating of both dark and bright soliton trains, depending on whether the carrier frequency lay on a region of negative or positive dispersion. The first demonstration of soliton-like spin waves due to induced modulation instability came in 1998 when Demidov [11] observed the effect in backward volume spin waves excited in YIG.

The ongoing interest of spin wave solitons [12, 13] signifies their technological importance and potential utility in a wave computing paradigm. Fundamentally however, they are also tied to another nonlinear phenomenon that is arguably more intriguing and that will become the focus of the next discussion.

---

[1] The Scottish engineer, John Scott Russell gives the first account of "that singular and beautiful phenomenon" in his 1844 *Report on Waves* [6]. He was observing a barge in the Union Canal which joins Glasgow to Edinburgh. When the barge suddenly stopped, the water around the prow formed "a rounded, smooth and well-defined heap of water which continued its course along the channel apparently without change of form or diminution of speed." Following this chance observation, Russell built a 30 foot wave tank in his garden and performed extensive experiments on the properties of solitary water waves.

5.1.2 Fractals

The study of fractals was solidified in the 1960s with Benoit B. Mandelbrot [14], and has captivated the interest of scientists ever since. Objects displaying fractal behaviour, or possessing a 'self-similarity', are remarkably plentiful in Nature, and exist in varied physical settings. Indeed, fractal analysis may be applied to systems including crystal growth, diffusion fronts, lung structure, moon-crater distribution, and stock market fluctuations [1, 15], and has even been employed in the authentication of the artwork of Jackson Pollock [16].

The ubiquity of these mathematical objects hints to us that there is something fundamental about them, while the simplicity with which they may be described is testament to their beauty.

Fractals may be classified using two main categories: exact and statistical. An exact fractal is an object that exhibits a perfect symmetry of scale: the structure of the whole is replicated upon varying degrees of magnification. This general condition can be applied to create some strikingly beautiful images like the Sierpinski gasket, or sculptures such as the fractal tree that may be found in Magdalen College, Oxford. Depicted in Fig. 5.1, each branch splits into two new branches, 45° to the original, resulting in smaller structures that replicate the whole.

A statistical fractal, is distinct from the exact fractal in that the shape is not replicated on different scales, but rather the statistical properties of the system are replicated. Perhaps the most famous example is the question asked in the title of Mandelbrot's seminal paper "How Long Is the Coast of Britain?" [14]. The answer is that the length of the coastline depends on the size of the ruler used to measure

Fig. 5.1 Fractal tree sculpture in Bat Willow Meadow, Magdalen College, Oxford. The structure emerging from each new branch mimics the whole on a smaller scale

it. As the scale of the ruler decreases, the length of the coastline behaves similarly to exact mathematical fractals. Perhaps intuitively, it is these statistical fractals that are the most common in Nature as they allow for random deviation from the perfect, exact regime which is very rare in the natural world.

Despite this rarity however, exact fractals have been observed in systems displaying nonlinear dynamics [17] and are closely linked with soliton behaviour. Space-domain soliton fractals were first simulated [18, 19] and experimentally demonstrated [20] in the field of nonlinear optics. In the field of magnonics, time-domain soliton fractals have been observed [21], driven by an active feedback ring, forcing them into existence.

Recently, spontaneous time-domain fractals have been observed in a magnonic system [22]. It was shown that spontaneous CW time-domain fractals can emerge in magnonic waveguides possessing highly nonlinear dispersion, as in the case of a 1D magnonic crystal. The experiment, performed by Richardson et al., utilised an artificial crystal with periodic grooves etched into it resulting in a rejection band for magnons with wavelengths corresponding to the periodicity of the grooves. The modified transmission band introduces a sharp kink in the dispersion curve [23]. The dispersion coefficient D then has a large spike at this frequency, resulting in an increased nonlinearity facilitating the generation of fractals when driven at high powers.

Crucially, the time-domain fractals observed in this experiment originated from self-modulational instability, where the fractal-like sidebands around the input frequency, represented amplitude modulation. These fractals appeared spontaneously, from a passive waveguide element, rather than being excited by an active feedback ring, for example.

In what follows, we present an experiment that builds upon previous observations of spin wave fractals and demonstrates spontaneous time-domain fractal behaviour of magnons that exist in an isotropic, unlithographed waveguide. This work contrasts with previous reports of fractals in magnetic media such that it neither exploits solitons, nor does it utilise a permanent artificial crystal. The precise mechanism for the observed effect is unclear, though the results are not inconsistent with observations from the literature [10, 21, 22, 24], which do not offer a consensus regarding the excitation mechanism.

5.2 Experiment

To create the fractal effect our experiment exploited a region of spatio-temporally periodic potential within the YIG waveguide. This region resembled a time-varying artificial crystal which we shall refer to as a dynamic artificial crystal (DAC). The DAC was created by the periodic potential associated with the presence of a standing wave across the width of the waveguide. From the well documented nonlinearity of YIG, the refractive index of the waveguide varies with the amplitude of the spin waves in a given region. When a standing wave exists across the width, the regions of

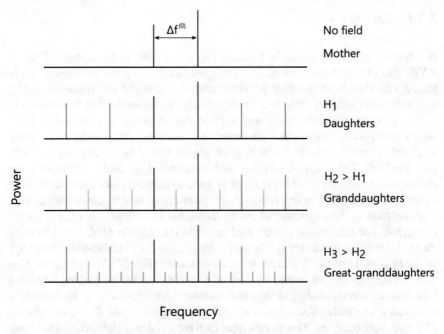

Fig. 5.2 Illustration of the observed fractal formation. Two inputs comprising the mother are detuned by $\Delta f^{(0)}$. As the field is increased to H_1, the comb of daughter modes, also with spacing $\Delta f^{(0)}$ appeared. At field H_2, a new comb of granddaughter peaks with spacing $\Delta f^{(0)}/2$ appeared. Finally, at H_3 great-granddaughter peaks appeared with spacing $\Delta f^{(0)}/4$

high amplitude are spatially fixed, forming a periodic refractive index, or an effective grating. This periodicity effects the transmission of certain wavelengths in the same way an etched artificial crystal does, significantly modifying the dispersion for the corresponding wavenumbers and frequencies and facilitating the emergence of fractal behaviour.

An illustration of the observed fractal behaviour is shown in Fig. 5.2. The effect behaves in the following way: (1) two input signals comprise the initiator (also known as the mother). (2) A comb of sidebands then appears which is referred to as the generator, or the daughters. (3) Half-way between consecutive peaks of the comb, a new set of granddaughter peaks arise. (4) Finally, this process is repeated and the emergence of great-granddaughter peaks was observed.[2]

---

[2]Some vagaries of the terminology used when describing fractal patterns should be clarified—the granddaughters and great-granddaughters may also be referred to as 1st pre-fractals and 2nd pre-fractals, respectively. The term *pre-fractal* is used when describing an exact fractal in the physical world. Since exact fractals are mathematical objects, they are infinitely scalable in a way that the physical world is not. This rather imprecise language allows us to keep track of which magnification level we are at and is conventional in the literature [18, 19, 21, 22].

5.2.1   Set-Up

A schematic of the set-up can be found in Fig. 5.3. Similar to the set-up in Chap. 4, a YIG thin film on a substrate of GGG was used as the magnon waveguide. It was placed YIG-side-down on a PCB with two pump antennae and a third antenna, acting as a transmitter-receiver. The same antenna design discussed in Sect. 3.3 was used.

The external magnetic field was applied out of plane to ensure FVMSWs were excited—a necessity since the experiment involves orthogonally propagating magnons. Figure 5.3 shows the waveguide region that is being pumped and supports the DAC. The potential oscillates with frequency f_{DAC}, and is one input of the mother mode. Also depicted in the figure is the transmitter-receiver antenna exciting a propagating FVMSW at frequency f', the other component of the mother mode.

A diagram of the experimental set-up including the instruments used is shown in Fig. 5.4. The microwave source used to excite the magnon DAC was a Hewlett Packard HP8672A frequency synthesizer, with $f_{DAC} = 3.915$ GHz and output power set to $P_{DAC} = 18$ dBm. A second microwave source (HP8671A) was set to $f' = 3.917$ GHz, also with an output a power of $P' = 18$ dBm before passing through a 9 dB attenuator and coupling to the transmitter-receiver antenna via a circulator. The circulator also enabled this antenna to send return signals to the Rhode & Schwarz ZVL spectrum analyser. The spectra were then recorded on a data acquisition computer and paired with the corresponding magnetic field as measured by the Hall probe.

The initiator of the fractal pattern is the mother mode, consisting of the two distinct frequencies, f_{DAC} and f'. The magnons that are excited at these frequencies also excited in separate regions of the waveguide. As discussed in Sect. 4.3.2, under the correct biasing conditions, the dispersion curve will satisfy the equation

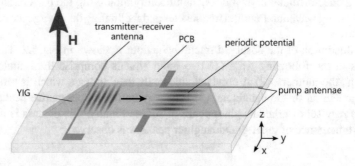

Fig. 5.3 Waveguide configuration: a YIG film on GGG substrate is placed YIG-side-down on a PCB designed with 3 antennae. The two pump antennae excite counter-propagating magnons which form a standing wave at specific magnetic fields. These create a time-varying periodic change in the refractive index due to the nonlinearity of the YIG, forming a dynamic artificial crystal (DAC). The transmitter-receiver antenna launches magnons towards this region and detects the response

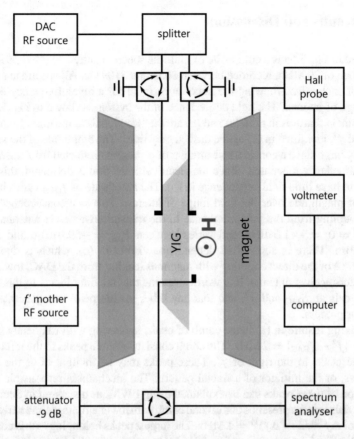

Fig. 5.4 Experimental set-up. The DAC microwave source is split and passed through isolators before exciting magnons to create the DAC. The second microwave source excites the other component of the mother mode at the transmitter-receiver antenna that is coupled to the spectrum analyser via a circulator. Spectra and magnetic field data are recorded on a computer

$$k_{\mathrm{DAC}}(\omega_{\mathrm{DAC}}, H) = \frac{\pi n}{w}, \tag{5.1}$$

where k_{DAC} is the wavenumber of the magnons that form the DAC, while $\omega_{\mathrm{DAC}} = 2\pi f_{\mathrm{DAC}}$, w is the width of the waveguide, and n is a natural number. As the incoming magnons with frequency f' enter the DAC region, the local distortion of the dispersion curve facilitates highly nonlinear phenomena [25], including the formation of a time-domain fractal.

5.3   Results and Discussion

Presented in Fig. 5.5a is a composite of multiple spectra centred on $f' = 3917$ MHz over a span of 20 MHz, recorded between 3052 and 3109 Oe. All spectra are plotted on top of each other, showing the maximum signal for a given frequency over the field range of interest. The field dependence of the system is shown in Fig. 5.5b.

Prominent features in both (a) and (b) are the two signals at the input frequencies f_{DAC} and f', identified in (a) by the dashed grey lines. The amplitude of the signal at f' is very large since a considerable amount of power was reflected from the antenna due to impedance mismatch. Since the electrical reflection is dominant, this signal appears to have little field dependence in (b). The amplitude of f_{DAC} varies by three orders of magnitude over the field range of interest. This is a consequence of the direct coupling from the pump antennae to the transmitter-receiver antenna which is $P'(H = 0) = -41.5$ dBm, and varies between $P'_{min} = -59.0$ dBm and $P'_{max} = -28.2$ dBm. There is some field dependence visible in (b), which is due to the interference of the direct coupling with magnons leaking from the DAC, towards the transmitter-receiver antenna. It is worth stressing that the only inputs to this system were signals at f_{DAC} and f', and that any other visible peak was generated by a nonlinear process.

A striking feature in (a) is the comb of peaks appearing with frequency spacing $\Delta f^{(0)} = |f' - f_{DAC}| = 2$ MHz. The comb consists of three peaks to the left of f_{DAC} and three peaks to the right of f'. These peaks may be thought of as the daughter modes, or the initiator of a fractal pattern. The mechanism responsible for the generation of these peaks can be explained using FWM, as discussed in Sect. 4.3.

Also visible is the presence of a second comb with lower amplitude and a frequency interval of $\Delta f^{(0)}/2 = \Delta f^{(1)} = 1$ MHz. The largest peaks belonging to this comb can be found in Fig. 5.5 occurring at $f = 3916$ MHz and $f = 3920$ MHz. This finer comb corresponds to granddaughter modes or the 1st pre-fractal. The physical origin of these granddaughter modes may also be explained with FWM. Consider the example of the granddaughter mode occuring at frequency $f_{DAC} + \Delta f^{(1)} = 3916$ MHz. Two magnons of this frequency can be obtained by the mixing of one magnon with frequency f' and one with frequency f_{DAC}. That is, $f' + f_{DAC} = 2(f_{DAC} + \Delta f^{(1)})$. This second, finer comb is similar to the phenomenon of period-doubling of solitons, reported in other nonlinear magnonic experiments [21, 26].

The field dependence of the combs is highlighted in Fig. 5.5b. It is clear that the comb of daughter peaks appears for a range of magnetic fields, one of which, H_1, is highlighted by the green dotted line. The granddaughter peaks however, have significant amplitudes over a much smaller field range, for example H_2, which is highlighted by the dotted blue line. The spectra for these particular fields are examined more closely in Fig. 5.6.

The top trace in Fig. 5.6 shows the inputs in isolation at zero applied field. These inputs constitute the mother mode or the generator of the fractal pattern. As the field is increased, side-bands at intervals of $\Delta f^{(0)} = 2$ MHz begin to appear. While the onset of these side-bands is not uniform a significant number are prominent at the

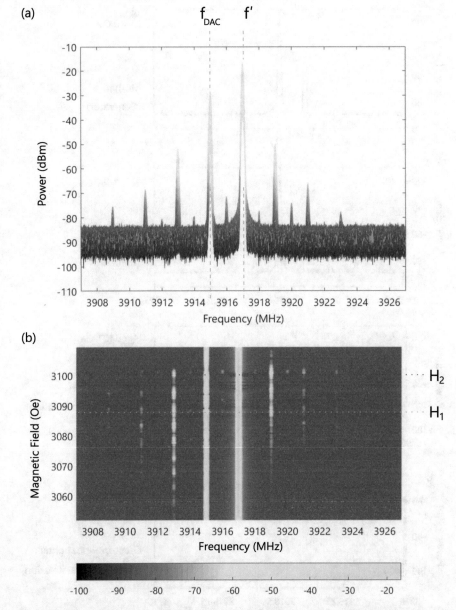

Fig. 5.5 a Shows a composite of multiple spectra, plotted on top of each other, demonstrating the maximum signal for a given frequency. The two input signals f_{DAC} and f' are indicated with dashed grey lines, while every other peak is due to a nonlinear response of the system. **b** Shows the spectra in (**a**) as a function of field applied magnetic field. Fields H_1 and H_2 are highlighted by the dotted green and blue lines, respectively. These spectra are shown in more detail in Fig. 5.6

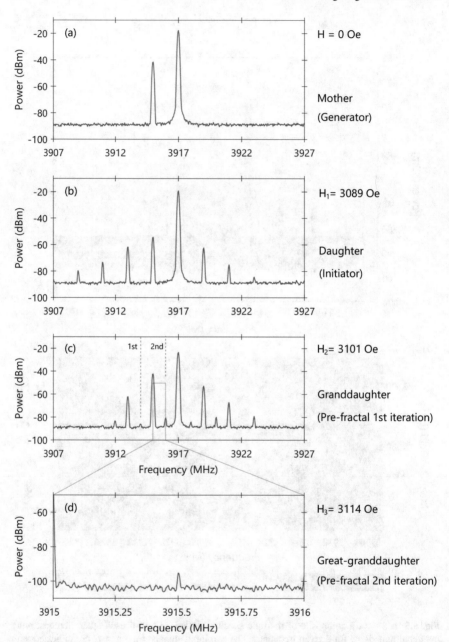

Fig. 5.6 Measured frequency spectra for various applied magnetic fields. Trace **a** shows the input signals constituting the mother mode (or generator mode) measured for zero field. **b** Frequency comb measured at $H = 3089$ Oe showing the daughter modes. **c** Presence of intermediary granddaughter peaks, midway between the daughter modes. 1st and 2nd granddaughter modes are indicated by orange dashed lines. **d** A magnified spectra, indicated by the (colour online) grey box and change of scale, showing the onset of a great-granddaughter peak

field $H_1 = 3089\,\text{Oe}$ as shown in Fig. 5.6b, where the comb begins at 3909 MHz, and can be seen clearly up to 3923 MHz.

As the magnetic field is increased further, granddaughter peaks become visible. These peaks were significant over a short field range, and were most prominent at $H_2 = 3101\,\text{Oe}$. These peaks have a frequency interval of $\Delta f^{(1)} = 1\,\text{MHz}$. It is clear that the granddaughter modes have a significantly lower amplitude than the daughter modes.

Since the amplitudes of two granddaughter peaks in particular are examined as a function of $\Delta f^{(0)}$ below in Fig. 5.9, the naming convention is emphasised here: the '1st granddaughter' shall refer to the signal occurring at $f_{\text{DAC}} - \Delta f^{(1)}$, while the '2nd granddaughter' shall refer to the signal at $f_{\text{DAC}} + \Delta f^{(1)}$. In the case of Fig. 5.6c for example, the 1st and 2nd granddaughters are indicated by the dashed orange lines at $f = 3914\,\text{MHz}$, $f = 3916\,\text{MHz}$, respectively.

The onset of a great-granddaughter comb is depicted in Fig. 5.6d. The trace is centered at $f = 3915.5\,\text{MHz}$, which is mid-way between the input at f_{DAC} and the 2nd granddaughter, as illustrated by the grey box. While (a)–(c) are taken from the same field sweep, (d) had to be recorded separately, and is a trace taken from the field sweep shown in Fig. 5.7. By performing a separate sweep over a narrower frequency span (1 MHz compared to 20 MHz), the resolution bandwidth could be decreased and therefore reduce the noise floor from approximately $-89\,\text{dBm}$ to $-103\,\text{dBm}$. By reducing the bandwidth in this way, the great-granddaughter peak with power $P_{\text{ggd}} = -95.3\,\text{dBm}$ was lifted above the noise.

Significantly, the peak in this trace appears with a characteristic frequency interval of $\Delta f^{(0)}/4 = \Delta f^{(1)}/2 = \Delta f^{(2)} = 0.5\,\text{MHz}$. The regularity with which the sidebands appear is commensurate with the formation of a time-domain fractal pattern as observed in a permanent magnonic crystal [22]. The pattern that we observed is pleasingly simple to express mathematically: the nth iteration pre-fractal should be a comb with frequency interval $\Delta f^{(n)} = \Delta f^{(0)}/2^n$.

The nonlinear peaks observed in Figs. 5.5 and 5.6 were extremely sensitive to changes in magnetic field strength. A case in point is the great-granddaughter peak occurring at $f = 3915.5\,\text{MHz}$. The field dependence of this signal can be found in Fig. 5.7, where the field ranges from 3111.9 to 3115.3 Oe. When the applied magnetic field was increased to $H_3 = 3113.7\,\text{Oe}$, a signal appeared at $f = 3915.5\,\text{MHz}$ corresponding to the 2nd iteration pre-fractal. There are two things to note about this great-granddaughter peak:

- The peak exists over a very limited field range $\Delta H \approx 0.1\,\text{Oe}$
- As the field is swept upwards, the formation of the peak is heralded by two separate peaks that combine to form the great-granddaughter peak.

The first of these observations led to some difficulty in measuring the signal, and indeed investigating the second observation further. This effect was however captured using the spectrum analyser in Fig. 5.8. The spectrum analyser is centred at $f = 3915.5\,\text{MHz}$ and set to a narrow frequency span of 20 kHz. The signals are not centred due to the spectrum analyser being imperfectly calibrated with a systematic frequency error of approximately 2 kHz. The purple arrows show the movement of

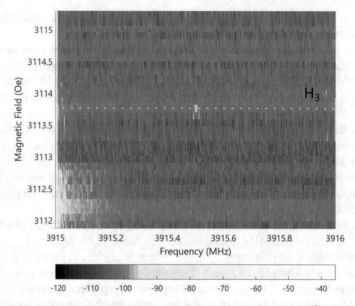

Fig. 5.7 Field dependence of great-granddaughter mode at $f = f_{DAC} + \Delta f^{(2)} = 3915.5\,\text{MHz}$. The dotted cyan line corresponds to the trace shown in Fig. 5.6d. As the magnetic field approached that at which the great-granddaughter peak was formed ($H_3 = 3133.7\,\text{Oe}$), two separate peaks (recorded in Fig. 5.8) appeared and then combined to create the 2nd iteration pre-fractal

the peaks as the field was increasing, during which time a snapshot of the signal was recorded on the spectrum analyser.

5.3.1 Detuning Effect

Having observed the existence of fractal behaviour above, the natural progression was to characterise their behaviour with respect to a controllable parameter. The experimental quantity that was chosen to vary was the detuning of the inputs that define the mother mode. That is, the frequency interval that defines the combs: $\Delta f^{(0)} = f' - f_{DAC}$. In varying $\Delta f^{(0)}$, f_{DAC} was held fixed at 3915 MHz, while f' was varied between 3916 MHz and 3925 MHz in steps of 1 MHz. For each value of $\Delta f^{(0)}$, two magnetic field sweeps were performed, whereby the spectrum analyser centred on the 1st granddaughter and the 2nd granddaughter (following the naming convention described in Fig. 5.6c). The frequency span of the spectrum analyser was set to 80 kHz to allow for a good signal to noise ratio. The noise floor for the inputs is approximately $-108\,\text{dBm}$, far below the signal power.

The data presented in Fig. 5.9 show the difference in power between the 1st and 2nd granddaughter fractals. Including the outlier, for which $\Delta f^{(0)} = 2\,\text{MHz}$, the mean absolute difference in power between the 1st and 2nd granddaughters is equal

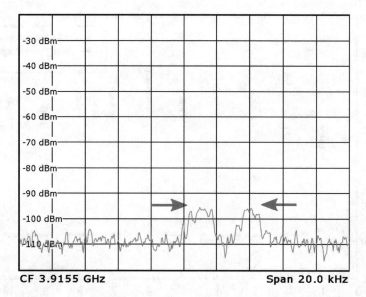

Fig. 5.8 Snapshot recorded on the Rhode & Schwarz ZVL spectrum analyser showing the presence of two peaks about to combine and form the great-granddaughter peak in Fig. 5.7

to 4.2 dB. Another feature of the data is that the lowest amplitude occurs with the smallest detuning. Indeed, when $\Delta f^{(0)} = 1$ MHz the power of the first granddaughter, $P_{gd_1} = -85.5$ dBm while the power of the 2nd granddaughter is $P_{gd_2} = -82.1$ dBm. This is consistent with the notion that in the limit where $\Delta f^{(0)} \rightarrow 0$ MHz, the granddaughters must disappear since they are no longer defined.

Discounting the 2 MHz outlier for which $P_{gd_2} = -65.3$ dBm, the maximum power for both granddaughters occurs at 6 MHz, where $P_{gd_1} = -67.0$ dBm while the power of the 2nd granddaughter is $P_{gd_2} = -70.2$ dBm. Over the measured range, no trend is immediately obvious within the data. Since the granddaughter modes are generated via spin wave FWM, the power of these modes will depend on the input power of the magnons. Therefore the changing efficiency of the transmitter-receiver antenna, shown in Fig. 3.16 may obscure any other underlying dependence on detuning.

The field dependence of the granddaughter peaks is also presented. Figure 5.10a–d show the amplitudes of the 1st and 2nd granddaughter modes for the input detuning values $\Delta f^{(0)} = 3$, 9 MHz as a function of magnetic field and frequency. The applied field was swept between 3073 → 3104 Oe for the 3 MHz detuning, and 3074 → 3105 Oe for the 9 MHz detuning. For each field sweep the spectrum analyser frequency was centred at the predicted granddaughter location, with each sweep spanning 80 kHz.

In each of the images (a)–(d) there is a central spine of signals that has some common features. An important feature is the intermittent nature of the spines with respect to magnetic field. Furthermore, this response to the magnetic field appears

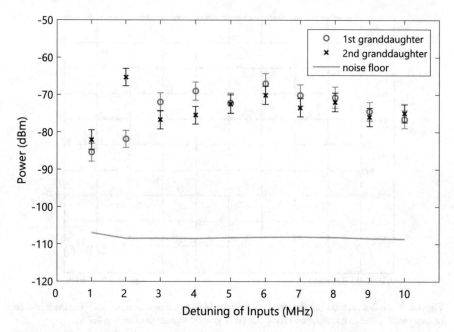

Fig. 5.9 Effect of detuning the inputs on the power of the 1st and 2nd granddaughter fractals. The maximum power of 1st and 2nd granddaughter peaks for a given $\Delta f^{(0)}$ is represented by the red circles and black crosses, respectively. The mean value of the background noise for each detuning is also plotted

to have a degree of periodicity to it. Figure 5.10e shows this field dependence of both granddaughters for an input detuning of $\Delta f^{(0)} = 3$ MHz, and similarly in (f) for $\Delta f^{(0)} = 9$ MHz. In both (e) and (f) there is a non-zero signal occurring approximately every 5 Oe. The significance of this period is discussed in Chap. 4. As shown in Sect. 4.2.1.2, this periodicity corresponds to the presence of a standing wave across the width of the waveguide; in other words, the formation of a DAC. This suggests that the fractal response of the system occurs only when a DAC is operational, which is not inconsistent with the observations of [22].

Another feature common to (a)–(d) worth remarking upon is the inconsistent spectral width of the spines. An illustrative example of this is the response around 3090 Oe in (c) and (d) which is much broader that the surrounding peaks. This response is symptomatic of the chaotic behaviour described in references [9, 27].

A different nonlinear effect is visible in (b), where there is an example of self-modulational instability. The peak occurring at 3085 Oe is preceded by two side-bands, approximately ±10 kHz from the central granddaughter frequency.

It should also be remarked upon that each of the subfigures (a)–(d) is offset from the predicted frequency by approximately +2 kHz. This is the same calibration error that was observed in Fig. 5.8.

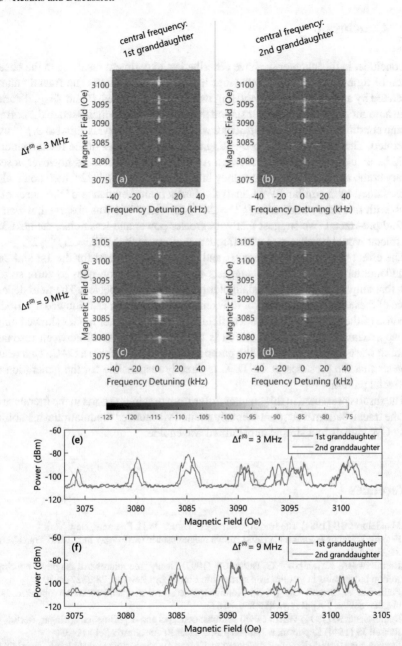

Fig. 5.10 a–d The field and frequency dependence of the of the 1st and 2nd granddaughter peaks, for an input detuning of 3 MHz and 9 MHz. The frequency span of the spectrum analyser is 80 kHz, with each trace centred on the predicted frequency at which the granddaughter mode should occur. Field dependences of both granddaughter modes for the $\Delta f^{(0)} = 3$ MHz (e) and $\Delta f^{(0)} = 9$ MHz (**f**)

5.4   Conclusion

To conclude, in this chapter we have described an experiment resulting in the observation of time-domain fractal behaviour in a magnonic system. The fractal pattern generated by a mother mode comprising two input frequencies f' and f_{DAC}, detuned by an amount $\Delta f^{(0)}$. It was demonstrated that as the field was increased to the correct biasing conditions, a comb-like structure with frequency intervals equal to $\Delta f^{(0)}$ was generated. These new peaks, representing the daughter modes of the fractal pattern, could be explained by FWM processes. As the field was increased however, a secondary comb appeared with a frequency interval $\Delta f^{(1)} = \Delta f^{(0)}/2$ which could also be explained by a similar FWM analysis. Furthermore, evidence of the onset of a comb with frequency interval $\Delta f^{(2)} = \Delta f^{(1)}/2 = \Delta f^{(0)}/4$ was observed, which is the 2nd pre-fractal. We suggest that with greater power and less noise, the next 3rd pre-fractal would be observed with frequency intervals of $\Delta f^{(3)} = \Delta f^{(0)}/2^3$.

The effect of detuning between f' and f_{DAC} was examined for the 1st and 2nd granddaughter peaks. Evidence suggests that as the detuning goes to zero, so too does that amplitude of the granddaughter peaks, as to be expected. The field dependence of these peaks, measured for different detuning values, also showed a periodic response of the system with respect to field. The granddaughter modes showed maxima approximately every 5 Oe, which is the same field interval between resonant standing wave modes excited by the pump antennae used to form a DAC. This result suggests that the presence of the DAC is directly responsible for the generation of the fractal pattern.

The results presented in this chapter differ from previous reports in the literature in that the fractal pattern was spontaneously excited via induced modulational instability using CW signals in a simple unpatterned waveguide.

References

1. Mandelbrot BB (1984) The fractal geometry of nature. W.H. Freeman, New York
2. Fetisov Y, Kabos P, Patton C (1998) Active magnetostatic wave delay line. IEEE Trans Magn 34:259–271
3. Kalinikos BA, Kovshikov NG, Patton CE (1997) Decay free microwave magnetic envelope soliton pulse trains in yttrium iron garnet thin films. Phys Rev Lett 78:2827–2830
4. Prabhakar A, Stancil DD (1999) Nonlinear microwave-magnetic resonator operated as a bistable device. J Appl Phys 85:4859–4861
5. Remoissenet M (1994) Waves called solitons: concepts and experiments. Springer, Berlin
6. Russell JS (1844) Report on waves. Rep 14th Meet Br Assoc Adv Sci 311–390
7. Serga AA et al (2005) Parametric generation of forward and phase-conjugated spin-wave bullets in magnetic films. Phys Rev Lett 94:167202
8. Zakharov VE, Ostrovsky LA (2009) Modulation instability: the beginning. Phys D Nonlinear Phenom 238:540–548
9. Kalinikos BA, Kovshikov NG, Slavin AN (1988) Envelope solitons and modulation instability of dipole-exchange magnetization waves in yttrium iron garnet films. Sov Phys - JETP 67:303–312

10. Wu M, Kalinikos BA, Patton CE (2004) Generation of dark and bright spin wave envelope soliton trains through self-modulational instability in magnetic films. Phys Rev Lett 93:157207

11. Demidov VE (1998) Induced modulation instability of spin waves in ferromagnetic films. J Exp Theor Phys Lett 68:869–873

12. Galkina EG, Ivanov BA (2018) Dynamic solitons in antiferromagnets (Review Article). Low Temp Phys 44:618–633

13. Sulymenko OR, Prokopenko OV, Tyberkevych VS, Slavin AN, Serga AA (2018) Bullets and droplets: two-dimensional spin-wave solitons in modern magnonics (Review Article). Low Temp Phys 44:602–617

14. Mandelbrot B (1967) How long is the coast of britain? Statistical self-similarity and fractional dimension. Science (80-.) 156:636–638

15. Addison P (1997) Fractals and chaos: an illustrated course. Institute of Physics Publishing, Bristol

16. Taylor R et al (2007) Authenticating Pollock paintings using fractal geometry. Pattern Recognit Lett 28:695–702

17. Segev M, Soljacic M, Dudley JM (2012) Fractal optics and beyond. Nat Photonics 6:209–210

18. Sears S, Soljacic M, Segev M, Krylov D, Bergman K (2000) Cantor set fractals from solitons. Phys Rev Lett 84:1902–1905

19. Soljacic M, Segev M, Menyuk CR (2000) Self-similarity and fractals in solitonsupporting systems. Phys Rev E 61:R1048–R1051

20. Erkintalo M et al (2012) Higher-order modulation instability in fiber optics. Int Conf Transparent Opt Netw 253901:14–18

21. Wu M, Kalinikos BA, Carr LD, Patton CE (2006) Observation of spin-wave soliton fractals in magnetic film active feedback rings. Phys Rev Lett 96:187202

22. Richardson D, Kalinikos BA, Carr LD, Wu M (2018) Spontaneous exact spin-wave fractals in magnonic crystals. Phys Rev Lett 121:107204

23. Ordóñez-Romero CL et al (2016) Mapping of spin wave propagation in a onedimensional magnonic crystal. J Appl Phys 120:043901 July

24. Cherkasskiĭ MA, Kovshikov NG, Kalinikos BA (2010) Observation of the modulation instability and spin wave solitons in ferromagnetic films under the conditions of coexistence of four-wave and three-wave parametric processes. Phys Solid State 52:2123–2128

25. Inglis A, Tock CJ, Gregg JF (2019) Indirect observation of phase conjugate magnons from non-degenerate four-wave mixing. SN Appl Sci 1:480

26. Gibson G, Jeffries C (1984) Observation of period doubling and chaos in spin-wave instabilities in yttrium iron garnet. Phys Rev A 29:811–818

27. Azevedo A, Rezende SM (1991) Controlling chaos in spin-wave instabilities. Phys Rev Lett 66:1342–1345

Chapter 6
Concluding Remarks

Throughout the preceeding five chapters, we have progressed from a general introduction of magnon-based computing and spin waves in magnetic films, to the detailed description of two specific experiments that investigate certain aspects of magnon behaviour.

It is hoped that this thesis has sufficiently expressed the results of the latter, and related them to their wider scientific and technological context. In this closing section, we comment briefly on the work as a whole and consider future prospects.

In Chap. 2 we introduced the central concept of reflections from a phase conjugate mirror. With focus on four-wave mixing—a particular method of creating a phase conjugate mirror—a simple theoretical model for phase conjugation in a magnonic system was presented by analogy with the well established optical model. From our model, a means of observing a phase conjugate spin wave was established.

A rigorous theoretical treatment of phase conjugation magnons via four-wave mixing is work for the future. Any prospective model should include damping, as this is an inescapable aspect of magnonics. Of particular relevance to the results in this thesis would be a magnonic theory of *non-degenerate* four-wave mixing, describing the reflection efficiency of a PCM that is oscillating at a different frequency from an incoming probe magnon. While the optical model provides a good insight, a magnonic approach would complete the picture and offer its own interesting physics to be exploited in a wave computing paradigm.

Detailed descriptions of the experimental background and apparatus were given in Chap. 3. Particular attention was paid to the antennae used in the excitation and detection of spin waves in the experiments of Chaps. 4 and 5. An antenna design with a meander structure that suppresses $k = 0$ modes and other long wavelength magnons was described, and a calculation of the magnetic field distribution for these antennae was performed to show that coupling to the $k = 0$ mode was sufficiently small.

© The Editor(s) (if applicable) and The Author(s), under exclusive license
to Springer Nature Switzerland AG 2020
A. Inglis, *Investigating a Phase Conjugate Mirror for Magnon-Based Computing*,
Springer Theses, https://doi.org/10.1007/978-3-030-49745-3_6

In Chap. 4 we reported on the evidence of phase conjugate magnons generated with four-wave mixing. We began with a description of the behaviour of and the conditions required to create a standing spin wave across the width of the YIG waveguide. A frequency response of the system was also shown, with a signal appearing at the expected phase conjugate frequency. Time lag measurements of the probe magnon return-trip time support the claim that phase conjugate magnons are generated in the pumped region of the waveguide. Simulations of the experiment that controlled for side-wall reflections were performed that also had a signal at the expected phase conjugate frequency, giving further credence to the claim that a phase conjugate spin wave was observed.

The challenges associated with directly observing the phase conjugate magnons were largely due to the impedance mismatch of the probe antenna. The reflection of the driving microwave signal at the antenna dominated the spectrum analyser reading requiring a non-degenerate approach. This in turn, destroyed the possibility of a direct comparison of the phase of the probe excitation signal and the phase conjugate reflection since the waves were no longer coherent. A perfectly impedance matched antenna—though difficult to manufacture—would have allowed the observation of a degenerate spin wave reflection from the PCM, since there would be no electrical reflection from the probe antenna at the same frequency.

While the simulations in Chap. 4 were useful in deducing that a phase conjugate spin wave was observed, there is scope for future improvement. An experimental set up that controlled for side-wall reflections—as was done for the simulations—is desirable. This may be achieved by evaporating a thin Au layer of a few nanometers at the waveguide edges to dramatically increase the spin wave damping. The limited access to YIG meant this action was not feasible, though it is encouraged for future experiments with greater resources.

The standing wave responsible for the phase conjugate magnons in Chap. 4, led to the unexpected nonlinear phenomenon that was presented in Chap. 5. The spatio-temporally periodic structure (the standing wave) combined with a monochromatic input (detuned by $\Delta f^{(0)}$ from the standing wave frequency) resulted in a time-domain fractal pattern. This is the first time (to the author's knowledge) that an exact fractal pattern has been observed via induced modulational instability in an unlithographed passive crystal. Furthermore, the observed pattern was fairly robust, and granddaughter fractal modes were observed over a range of detuning frequencies.

Again, better impedance matching of the transmitter-receiver antenna would offer further avenues for measurement. In particular, a time-domain measurement with a sensitive oscilloscope would allow one to investigate whether or not the fractal pattern observed in the spectra corresponded to a soliton pulse train. If this were the case, a soliton train with tunable period would be observed, given the detuning range over which fractals appeared. Such a phenomenon might be exploited in wave computing as a means of carrying information.

In this work, a specific advancement of magnonics has been achieved, in the demonstration of phase conjugation via four-wave mixing. The realisation of this process in a spin wave system opens the door to the implementation of a phase conjugation operation in a computational setting that exploits spin waves. Our experiment has demonstrated that another tool is available in the expanding magnonicians tool-box.

Central to this work was the presence of the spatio-temporally periodic potential created by the standing wave, or the dynamic artificial crystal. The nonlinear phenomena reliant upon this—both expected (Chap. 4) and unexpected (Chap. 5)— suggest that there may exist other uses for such a system; new and fundamental spin wave physics, and potential spin wave devices may be discovered with further investigation of this interesting dynamic periodic potential.

With the need for improvement of computational technology showing no signs of slowing, there remains a demand for a low-power alternative to work in tandem with CMOS. To this end, the door is open for magnonics to enter. A number of challenges remain however, before spin waves can be widely adopted by the computing industry. From an engineering standpoint, there is the difficulty that for any device exploiting spin waves, there must be a considerable magnetic bias field present to support and tune the spin waves. It was also explored in Chap. 1 that the greatest rewards of magnon computing—smallest wavelengths and highest frequencies— may be reaped using *exchange* spin waves. Since the vast majority of modern spin wave research concerns dipolar regime, there are many aspects of exchange waves that remain to be investigated. Indeed, a lot of interesting spin wave physics arises precisely because of the novel dispersion curves of dipolar spin waves, specifically backwards volume magnetostatic spin waves. Many of these useful effects (such as caustics) are simply not present with exchange waves.

A significant hurdle for magnonics in general is developing an efficient method for exciting and detecting spin waves. The methods used at present are sufficient for research, but a marketable solution remains currently out of reach. This fact is truer still for exchange waves, where there is greater difficulty in excitation and detection. Furthermore, assuming incremental improvements in the excitation of smaller wavelengths, the benefits of exchange waves can only be unlocked by first traversing the comparably flat region of the dispersion that comes with dipole-exchange waves. The flatness of the dispersion means that these waves propagate relatively slowly, and in some cases are entirely stationary, offering no utility for data transmission.

Despite these difficulties, spin wave computing remains a strong contender for a complementary technology to CMOS. Whether or not this technological gap will be filled by magnon-based computing, we cannot say. The richness, vibrancy, and beauty of spin waves however, is certain.

Curriculum Vitae

Alistair Inglis

Publications Related to This Thesis

Alistair Inglis, Calvin J. Tock and John F. Gregg. Indirect observation of phase conjugate magnons from non-degenerate four-wave mixing. *SN Appl. Sci.* **1,** 480 (May 2019).

Alistair Inglis, John F. Gregg. Onset of spin wave time-domain fractals in a dynamic artificial crystal. *J. Magn. Magn. Mater.* **495,** 165868 (Feb 2020).
John F. Gregg, Burkard Hillebrands, **Alistair Inglis**, Monika E. Mycroft, Evangelos Th. Papaioannou, James Semple, Calvin J. Tock. A Microscopic Explanation of Microwave Spin Pumping. *Submitted.* https://arxiv.org/abs/1711.06048

Education

DPhil, University of Oxford, Condensed Matter Physics, 2015–2019
MSc, University of Glasgow, Physics (1st class), 2010–2015
Syracuse Scholar, Syracuse University, New York, USA, 2009–2010

Experience

Data Scientist, Department for International Development, 2020–Present
Tutor, Greene's Tutorial College, 2018–2019
Lecturer, University of Oxford, 2016–2017
Research Scientist, Diamond Light Source, 2014
Research Scientist, University of Glasgow, 2013
Programmer, Lawrence Berkeley National Laboratory, USA, 2012–2013

Awards and Prizes

Syracuse-Lockerbie Scholar (Lockerbie and Syracuse University Scholarship Trust)
Beghian Scholarship for Physics (Magdalen College, University of Oxford)
Mackay Smith Prize for 1st in year for Physics (University of Glasgow)
Merit Award (University of Glasgow)
Travel grants for research and education (Glasgow, Oxford, Kaiserslautern)

Printed in the United States
by Baker & Taylor Publisher Services